TRIVENT
PUBLISHING

TRIVENT

TRANSHUMANISM

Violent Tech:
A Philosophical and Theological Reflection

Joshua K. Smith

Ethics and Robotics, vol. 3

Violent Tech: A Philosophical and Theological Reflection
Joshua K. Smith
ISBN 978-615-6405-83-8
Ethics and Robotics, vol. 3
ISSN 2939-5992
Responsible publisher: Teodora C. Artimon
https://trivent-publishing.eu/76-ethics-and-robotics

First published in 2023 by Trivent Publishing

Trivent Publishing
1119 Budapest, Etele ut 59–61
Hungary

For more information, visit our website: http://trivent-publishing.eu

Compliance

We just need your compliance
You will feel no pain anymore
And no more defiance
We just need your compliance
Just give us your compliance
We won't let you feel lost anymore

No more self-reliance
(Com-com-com-com-compliance)
Fall into line, you will do as you're told
No choice fatigue, your blood is running cold
We lose control, the world will fall apart
Love of your life will mend your broken heart
Life lived in fear, you need protection

You're all alone, too much rejection
We have what you need, just reach out and touch
We can save you (We can save you), we just need your
Cover your tracks, we know what's best for you
You've bitten off much more than you can chew
You're running scared, you'll run into our arms
Come join our clique, we'll keep you safe from harm

Our toy soldier, you'll do the dirty work
Stay loyal to us, we'll take away the hurt
We have what you need, just reach out and touch
We can save you (We can save you), just give us your

Muse, "Compliance," *Will of the People*, 2022

TABLE OF CONTENTS

FOREWORD

By Elke Schwarz

It is a pleasure to write the foreword to this powerful book. That is perhaps a strange thing to say about a text with such a grim subject matter. Speaking about violence is not an easy task. It is difficult to address openly and frankly what we often would rather pretend not to see or know about. To do so in the context of both religion and technology is an even more challenging endeavor, but one that is sorely needed in a world marred by technological modes of destruction that could very well wipe out the last shreds of humanity. This book is an important call to not avert one's eyes. It presents a clear and unflinching look at the violent substrate of our lives. This is what makes *Violent Tech* such a daring book. It is also a timely book, because we are at a juncture where technologies of violence increasingly appear to subsume the need for justifications of violence and, indeed, normalise greater use of violence through the veil of efficiency and neutrality. More and more, we are losing the space to contest the ostensible justness of lethal technology, and we are losing sight of the fact that violence is always a choice for which someone must take responsibility. Although the book takes a theological perspective, *Violent Tech* makes a strong case for human agency. We have a choice in how we wish to proceed with our technological innovations: continue to fold them into military endeavours, even though this will likely bring about more and not less violence; or else make a conscious and concerted effort to design technologies that positively cultivate the relationships we have with others, as well as with ourselves.

The book takes us through a conceptually rich theological reading of its key concepts (violence and technology), underwritten by an impressive command of a wide range of literatures. I particularly welcome how the text foregrounds the thought and insights of women scholars. This is rare, but absolutely necessary, for a book on violence. The business of war, destruction, and violence is all too often considered to be the purview of men, whether as politicians, perpetrators, or analysts, and more so still as violence and technology become closely entwined. Indeed, as many have highlighted, the very notion of technology is typically ascribed to the domain of male activity. The couplet 'man and machine' sadly still rolls off the tongue as easily as 'military and masculinity'. This leaves the debate on violent technologies single-focused and anemic. This is despite many, many women scholars having tirelessly raised their voices as credible analysts, experts, and critics of violence in past centuries. The approach Joshua Smith offers in this book opens perspectives and viewpoints up to these voices, giving the discussions in this text a particular depth and strength, and that is so very important.

With this firm grounding in place, the book examines how the unfettered advancement of death machines has become wired into the contemporary psyche. Death machines are cast as exciting innovations; they are insidiously folded into gaming and other modes of entertainment; they represent the manifestation of power in all its dimensions. Often occluded in this casting is how they "violate personal space, privacy, body and even soul as they determine and alter how the human team member operates", as Joshua Smith points out in the text. The goal, the telos, of any violent technology, he says, is the exploitation of the human. This observation is spot-on and forms the basis for what I take to be the book's core critique: that the pursuit of violent technology is a pathway to the complete dehumanization of both those targeted and those applying the violence.

Technologies of violence are not mere instruments; they shape norms and normalize violent practices; they create rifts between us, but cannot create new positive or indeed peaceful contexts. As Hannah Arendt already noted in her 1970 essay *On Violence*, violence is instrumental, it is means-ends oriented. In fact, it relies to a considerable degree on instrumental implements, tools, and technologies for its execution. Violence can destroy power. But power cannot grow out of violence. Violence cannot promote causes, or make a different future. By extension, neither can violent technologies. Yet the casting of new Artificial Intelligence (AI) enabled weapons as somehow ethically superior, or as bestowing some form of exquisite power which is so prominent in military discourses today, suggests exactly that – that violence can "make" something better. Following Arendt, violence as a creative force serving the notion of "making" a human history is not possible. Violence breaks, it injures, it ends things. It does not create – it is the opposite of new beginnings, the opposite of fostering relations. If violence is used to bring about a new relationship of care or peace, it will fail, as such interpersonal outcomes cannot be secured through violence. Joshua Smith's thoughtful and rigorous book helps us better understand just how misguided current views on the potential of AI weapons are!

Yet this is also a book of hope; hope that by looking unflinchingly at the violence-technology-theology nexus, we can reveal the myths that underwrite the addiction to violence through technology that marks our modern existence, and, importantly, "think otherwise". The book argues that we can do so, and that we have a duty to our fellow creatures to do so. In this vein, we must focus on making technologies only with positive aims in mind ("gardens, not guns"). Moreover, we should recognize a duty to be more attentive to the machines we are making in terms of their moral worth. I, myself, would perhaps not quite take quite such an intrepid leap into extending moral ontologies, but I understand this move in the context of a deep concern with ethics, and with this I sympathize.

I am certainly a fellow traveler in trying to think the world as something other than intrinsically violent, in thinking that we can be otherwise, and that we have an ethical responsibility to our fellow creatures to do so. But everything depends on our willingness to do so. At a recent high-level summit on responsible military Artificial Intelligence, one of the most frequently used metaphors by delegates was that 'the train has left the station' – we have AI weapons and so we should just focus now on using them responsibly. Very few voices were heard pushing against this strange self-fulfilling prophecy of runaway violent technologies. But there is a way to think against this grain and *Violent Tech* is a wonderful resource for doing so.

I have no doubt that everyone reading this book will come away from it a richer person.

INTRODUCTION

This is a dangerous book. There were several moments in the process of reflecting on both texts of violence and technology where I contemplated abandoning the project altogether. Since the beginning of this project, I have suffered many episodes of anxiety, moral injury, and depression. At the beginning I would caution the reader to continue at their own risk. I serve as a pastor in a rural part of Mississippi. On a weekly account my job requires me to walk alongside people in their trauma, hurt, grief, and pursuit of reconciliation with the violence of this world. Because of this reality, there are moments of pessimism in my writing and what I believe about the future of our relationship with technology. All hope is not lost. Education is our pathway to freedom, and this is certainly the case in regard to understanding those that promote the use of violent technologies. There is a healthy fear of AI and robots evidenced in the lineage of science fiction. How poetic that a female writer and scholar such as Mary Shelley warned modernity of the dangers of creating implements without ethical consideration. I say this is poetic because much of modern technology we owe to the forethought of the female mind. Chemistry and medicine are rooted in cooking, architecture is rooted in weaving, for a few examples. I am not sure that anyone understands violence and technology the way that women do. Nor do I believe that there is a shred of significant technology that is not linked to the female mind.

I would like to take this opportunity at the forefront to acknowledge the unfortunate hiddenness of women in holding up beneficial and ethical technology. I am grateful for the edited work of Joan Rothschild *Machina ex Dea*[1] and the various feminist perspectives on technology that it brought to my attention. I am also deeply indebted to the scholarship that preceded me, many of whom are female. In the long survey of literature that surrounds both violence and technology, I noticed a different authorial voice between men and women. Specifically, the female voices, especially those of minorities, had a more holistic perspective of both the political nature of violence and technology and how such reverberates throughout generations. I do not think this is a coincidence at all. Rather, it is informed by the mental and physical experience of the female body throughout history.

I struggle to think about violence and warfare without a womanist framework.[2] Works like Elaine Brown Crawford's *Hope in the Holler* frame the reality of violence and its mediation through technology astutely. Yet, there is a continual voice of hope and resistance. Womanist theology is opposed to the idealized future visions of frameworks like Jürgen Moltmann's theology of hope which orients the perspective toward the future. Thus, scholars like G. Clarke Chapman and James Cone have criticized this framing for ignoring the impossibility of this future for people of color.[3] Voices like Cheryl A. Kirk-Duggan in her work *Violence and Theology*[4] examines the pandemic of violence in the late 80s and 90s toward women and children, offering theological reflections on how to move forward. I also think about the work of Bell Hooks and the issue of self-love and identity that many black communities suffered both from physical rape, beatings, and torture, as well as psychological, for

[1] Joan Rothschild, eds. *Machina ex Dea: Feminist Perspectives on Technology*, The Athene Series (New York: Pergamon Press, 1983).

[2] This term originates from the work of Alice Walker's *In Search of Our Mother's Garden: Womanist Prose* (San Diego: Harcourt Brace Jovanovich, 1983). It stems from the work *womanish* in reference to African American woman or Black Feminism. It is a literary means of understanding the oppression of women of color.

[3] G. Clark Chapman, "Black Theology and Theology of Hope: What Have They to Say to Each Other?" in *Black Theology: A Documentary History, 1966–1979* (Maryknoll, NY: Orbis Books, 1979), 193–215; James H. Cone, *God of the Oppressed* (New York: Seabury Press, 1988), 126–32.

[4] Cheryl A. Kirk-Duggan, *Violence and Theology*, Horizons in Theology (Nashville, TN: Abingdon Press, 2006).

example, being taught to hate their phenotype. As the reader will see in this book, I mention womanist theology for a couple of reasons. While this is not a work of womanist theology, that is, it does not solely focus on a womanist hermeneutic or reading of African American experience strictly for ethical formulation; I have found the focus on the victim-survivor experience helpful in my theological and philosophical reflection. For as the brilliant works of Ruha Benjamin, Cathy O'Neil, and Safiya Umoja Noble[5] have also taught us, violence does not end with ideology; instead, it continues in our technology. No, this is not a womanist critique of violence and technology, but it does share the concerns of said approach. I cannot faithfully articulate and analyze the experience of violent tech upon the perspective of a woman, but I can share its concern to find pathways to healing, liberation, and hopefully transformation.[6] Although I am not claiming that this work is one of womanist theology or ethics, this work is concerned with the marginalization of the powerless and the violation of persons (emotional and physical). It is in this spirit of overcoming adversity that I seek to conclude each chapter, asking the question: Where are justice and hope amid death, destruction, and dehumanization employed in violent tech?

As a pastor-theologian, I must be a voice of hope amid technological pessimism. Too many critiques of technologies, especially from a Christian perspective, promote a luddite approach that neither deals with practical solutions nor promotes hope for our relationship to technological innovation. A trajectory of hope begins with an ethical reading of why violence occurs and why implements inevitably become our partners in crime. Reading as an ethical project gives me hope, but also holds me to realism, avoiding the pitfalls of what has become known today as *Longtermism*: an ethical framework that gives priority to the future. I do not think this stance is anything new, rather it is what I, and others, label as *Robotic Futurism*: placing hope in a better future in the

---

[5] Ruha Benjamin, *Race After Technology: Abolitionist Tools for the New Jim Code* (New York: Polity Press, 2019); Cathy O'Neil, *Weapons of Math Destruction: How Big Data Increases Inequality and Threatens Democracy* (New York: Broadway Books, 2017); and Safiya Umoja Noble, *Algorithms of Oppression: How Search Engines Reinforce Racism* (New York: New York University Press, 2018).

[6] Toinette M. Eugene, "...If You Get There Before I Do...!": A Womanist Ethical Response to Sexual Violence and Abuse," in *Perspectives on Womanist Theology*, ed. Jacquelyn Grant (Atlanta: GA: ITC Press, 1995), 95.

hands of technological implements. In these ethical frameworks, most, if not all, of our current political and economic circumstances will dissolve (somehow) by future innovation. That is not to say that those who hold these views do not value present ethical dilemmas, but that they equally view the future of society as a concern in their ethical considerations.[7]

Reading is not only an ethical project but also a moral formation. Through encountering the text of the Bible, its stories of violence, and the modern justification of death machines, the reader will see the aid of narrative ethics. While there are many authors that can aid the reader here, my choice in this work bends toward the Levinasian influence. Why Emmanuel Levinas? In dealing with the heaviness and ethical problems around nebulous topics like violence and technology; Levinas is of great importance because of his emphasis on alterity and responsibility. Understanding the ontology of violent implements, from a Levinasian perspective, is not the primary focus of ethical philosophy. What is learned about ethical responsibility through the limited understanding of the ontological difference between humans and machines? Per the discussion in chapter 5, not much is answered in this understanding, nor does it solve all of the problems facing their use in warfare and policing.

While Levinas can be used as a guide for reading the biblical text, I seek to bring in his ethical framings more so on the philosophical side of our consideration of violent implements. For the hermeneutical lens employed in chapter 2 (and most other theological sections), I will employ a narrative ethic that resembles the work of Amy C. Cottrill per her work on biblical violence, *Uncovering Violence: Reading Biblical Narratives as an Ethical Project.* [8] Likewise, Corttill uses both a Levinasian approach and the hermeneutical lens of Martha Nussaum and Adam Zachery Newton. The purpose of reading the narratives this way is to do ethics, rather than merely extracting an application for modernity. It is, as Cottrill writes, "recognizing the alterity of the biblical world and the biblical texts as well as their force and infectious potential."[9] The bible is not an instrumental technology, but an invitation to the suffering and experiences of the world it reflects in conversation with the reader.

[7] William MacAskill, "The Case for Longtermism," *New York Times*, August 7, 2002, https://www.nytimes.com/2022/08/05/opinion/the-case-for-longtermism.html.

[8] Amy C. Cottrill, *Uncovering Violence: Reading Biblical Narratives as an Ethical Project* (Louisville, KY: WJK Press, 2021).

[9] Cottrill, *Uncovering Violence*, 39.

Cottrill gives some helpful principles that I consider in my reading of narrative ethics:

> (1) Narrative ethics is attentive to the aesthetic shape, form, and particularity of the text as an invitation to otherness. (2) Narrative ethics resist summative claims or ethical prescriptions … (3) Narrative ethics assumes an understanding of the self that is porous and responsive to the other, including the text. (4) Narrative ethics assumes that what is felt within the bodies of readers is valuable and formative knowledge about violence and its effects on those bodily impulses and sensation shape and construct the cognitive landscape in which conscious awareness and reasoned reflection occur … (5)Narrative ethics affords a way of reading texts that focuses on what one can see, touch, and be accountable to as a result of being in proximity to the other, or to the text; this mode of ethics is distinct from one that priorities articulation of ethical values, principles, or conclusions.[10]

By using a variety of ethical frameworks to read and interpret, I hope to provide the reader with a wider perspective on how to understand violence and its relation to technologies of death and harm. It is impossible to assume that one monograph on violent technology will solve all the problems of reading difficult text and how it relates to the struggle for peace and freedom from oppression. I hope and pray this work is one that further promotes peacemaking, reconciliation, and forgiveness in the anthropological struggle for security and significance.

Chapter 1 will survey the terminology of violence, technology, and what happens when the two merge into one. In understanding the metaphysical and religious undertones of technology, one can see how violence links to a vast number of implements. Here, it is also necessary to explore Christian perspectives of violence and the fall, indicating the connection between the legacy of violence and the desires humanity has held that technology would bring about. In chapter 2, the reader will examine the alterity of human violence through three lenses: the episode

[10] Cottrill, *Uncovering Violence*, 39–40.

of Cain and Abel, war texts, and the non-retaliation of Jesus. Here we will also explore the theological implications of violent tech and the theological vision of hope.

The bad romance between DARPA and the Department of Defense will be the focus of Chapter 3. This book will assess violent technologies such as autonomous weapons systems and the ethical, ecological, and economic costs behind these systems. The US military-industrial complex's addiction to war-making is a theological as well as a philosophical issue for both the consumer and combatant. While robots designed for the theatres of war are modern conceptions, the desire for security and economic stability is an ancient one. The justification of violence through warfare is based on both political and theological ideologies. Chapter 4 will explore the use of violent video games by both civilians and militaries, and the narrative it communicates about violence. In addition, this chapter will cover the moral injury and potential of redemption of these technologies for such technology.

Chapter 5 will critique the use of violent tech, especially lethal autonomous weapons, in light of just war theory. In this chapter, I posit the ethical and economic cost of these implements and provide some practical recommendations, per our Levinasian and narrative ethic lens, about our future among such desires for implements of war.

CHAPTER 1

Violence + Technology = Violent Technology

The terms *violence* and *technology* are as broad as they are ambiguous. This reality makes both terms elastic, and perhaps this is for a good reason. Authors and researchers pull and stretch these terms to mold them to a particular meaning that is in a particular discussion. Regrettably, there is an application of *psychological essentialism* involved in this framing of terms. Because there is ambiguity within the terminology, one can operationalize the terms to fit their ideological, religious, and philosophical context. The irony here is that this is a sort of *violence* that is being imposed upon the very words meant to conceptualize the subject under discussion. The history of interpretation of the Bible alone is witness to this phenomenon. While that may be, the reader should not lose heart, for there is a large body of literature to help guide us on this journey to understand violence and its implements.

How do we find the *true* meaning of such eclectic terms? One way is the use of heuristics, and whether that is for restoration or for ruin depends on the cognitive abilities of the interlocutor and the informational fidelity of the resources at hand. We cannot exegete technology without using a *technique*, nor can we faithfully examine a piece of technology in isolation from a worldview that is situated in a world of violence and violation. Two important works help the modern reader in formulating technology: Martin Heidegger's essay, "The Question

Concerning Technology"[1] and Marshall McLuhan's work in *Understanding Media: The Extensions of Man*.[2] McLuhan's thesis argued that technology extends oneself and thus technology has a lateral effect on society. This reading of technology as a mere tool or "extension" is important, but it is only half of the story, so to speak. McLuhan is right in the sense that technology, through the combination of materials, reemerges into new *media* of slight but significant change upon and within the operators' ethical frameworks.

McLuhan's view made room for the most disparate places found within discussion of technology. For the religious scholar, one can quickly locate McLuhan's extension view as an adaption of Thomas Aquinas' work on formal causality. It is important to note here that McLuhan (later in his life) was Catholic. Remember, one's worldview shapes the hermeneutical process for understanding any technology and the role the self plays within it. In a letter to Fr. John Culkin, McLuhan wrote,

> My own approach to the media has been entirely from formal cause. Since formal causes are hidden and environmental, they exert their structural pressure by interval and interface with whatever is in their environmental territory. Formal cause[s] are always hidden, whereas the things upon which they act are visible.[3]

We will return to McLuhan's theory momentarily; however, first we need to clarify what he means by formal causes. For those not familiar with the scholastic metaphysics mentioned above, let me give a brief primer. Since the 1700s there has been a revulsion to the binary conceptualization of Aristotelian metaphysics which is recapitulated in

---

[1] Martin Heidegger, "The Question Concerning Technology," in *The Question Concerning Technology and Other Essays*, trans. William Lovitt (New York: Harper Perennial Modern Thought, 2013), 3–35.

[2] Marshal McLuhan, *Understanding Media: The Extension of Man* (Cambridge, MA: MIT Press, 1994).

[3] Marshall McLuhan, *The Medium and the Light*, eds. Eric McLuhan and Jacek Szkiarek (Eugene, OR: Wipf & Stock, 1999), xx–xxi.

the work of Thomas Aquinas during the scholastic era (and rightly so).[4] When I say binary, I mean the general distinction between what an object *is* and what it *can* become. In the language of philosophy, we know this as the theory of act and potency. Or, as Bernard Wuellner summarizes the Thomistic rendering, "Potency and act are a complete division of being. Hence, whatever is must be either pure act or a unit composed of potency and act as its primary and intrinsic principles."[5]

The first of the twenty-four theses from Thomistic thought shows the fundamental distinction between act and potency in scholastic thought. When it comes to metaphysics (i.e., substance, essence, and causation), the details matter. The details that matter most at present are that, in Aristotelian-Thomism, there is a real distinction between what things/persons are in act and what they are in potency. On one side of the spectrum with thinkers like Parmenides and Zeno, there is no distinction and no reality of change–it is illusionary. On the opposite end of the spectrum, in Heraclitan thought, there are no intrinsic principles, no laws, or biological species; however, all things are perceptual experience. Aristotelian-Thomism is a middle-ground or complexity of these extremes. Why is this problematic in the 1700s specifically?

To approach this question is to look at the core of what modern phenomenology has sought to respond to. The Aristotelian binaries and categories are the foundation of modern biological and physical sciences, yet since Descartes there has been a rebellion–dare I say violence towards this logic. The rejection of Aristotelian metaphysics is in essence a rejection of the ultimate cause, which is God. But there is also a categorical error being made in this logic. Aristotle, in my reading, does not drive complicated binaries between the philosophy of science and metaphysics. In fact, the theory of actuality and potency is applied in both Aristotle's philosophy of nature and metaphysics. While there are disagreements about this view, I am sympathetic to the view of Andrew Van Melsen's work on this matter, which reads that the Aristotelian-

[4] The political world of both Aristotle and Aquinas disregarded the rights of women and children and saw many humans as less than. No historical figure is free of ideologies that encapsulate their era of time. It is the ethical duty of the reader to extract what is ethically and morally appropriate for their era, knowing that ultimately it will be considered mired by the next.

[5] Bernard Wuellner, *Summary of Scholastic Principles* (Chicago: Loyola University, 1956), 120.

Thomistic view of the philosophy of nature is a branch of metaphysics.[6] This assumption is important for a couple of reasons. First, this assumption holds that all potential change is grounded in prime matter. A simple statement with huge implications. Famously, David Hume challenged this principle by theorizing that something can come into being without cause. Conceptualizing is not the same as conformation through empirical observation. Even in evolutionary theory, change is never from *nowhere* for genetic mutation, changes in the environment, and selection leads to effectual change. An object's being determines its final causality. For example, naturally, the final cause of a pine seed is to become a pine tree. There are effects that can alter or deter this end for coming to be, however, that does not falsify the reality of the natural effect. In materialistic philosophy, the previous statement is offensive and mired (i.e., debates around emerging gender identities).

Second, there is a distinction between natural objects (i.e., substantial forms) and intrinsic teleology. In the mechanical worldview, following Descartes' mechanics of nature, matter is the extension of space. In this view matter is atomized and no longer the potential of form. This view is a radical rejection of formal cause. The gravitation toward Parmenidean through developments in Newtonian theory, electromagnetism, and quantum mechanics, is just as much a metaphysical and epistemological assumption as saying that God is the prime mover. In fact, much of the disgust with Aristotelian metaphysics stems from a desire to ban purpose or final cause (more to come on that later). In the words of Alex Rosenberg,

> Ever since physics hit its stride with Newton, it has excluded purposes, goals, ends, or designs in nature. It firmly bans all explanation that are teleological...No matter how physics eventually deals with these problems...[i]n solving them, physics will...not give up the ban on purpose or design.[7]

It is strange to see this type of fundamentalism within secular philosophers, but it is revealing. Just because a modern philosophy of nature "bans" purpose that does not empirically prove that purpose does

[6] Andrew Van Melsen, *The Philosophy of Nature*, 2nd ed. (Pittsburgh: Duquesne University Press, 1954), 98.

[7] Alex Rosenberg, *The Atheist's Guide to Reality: Enjoying Life without Illusions* (New York: W. W. Norton and Co., 2011), 40–41.

not exist. Okay, but what does this have to do with McLuhan's view of formal causes?

Formal cause correlates to the teleology of an object or person. For McLuhan this refers to the potency of a piece of technology and the metaphysical assumptions of the creator of said technology. Marshall and Eric McLuhan's joint work *The Laws of Media* was an in-depth remix of Aristotelian causality. After reading the brief remarks above about actuality and potency, one can see why McLuhan emphasizes formal causes in the realm of technological critique. There is an ideology that undergirds the tools we create to extend ourselves. In the realm of media, McLuhan is not giving ethical commentary on a given piece of technology, but rather acknowledging the external pressure that a form of technology acts upon society. The concern for the individual self is what is of utmost importance for the reader, no? This question is a reverberation of the human obsession with efficient causation.

More to this, the disappearance of formal causality in philosophy with Descartes in the 1700s is juxtaposed to its appearance in the media of art during the romantics. What the romantic authors and McLuhan conclude about new media and technology is that it is *forms*. The creator of said forms has no control over the external effects. The human interlocuter is passive and disadvantaged towards the *force* of the media. Now we are getting closer to the essence of violence and technology.

What Is Violence?

Even the most careful anthologies, like the one edited by Vittorio Bufacchi and David Riches, will attest any study of violence is exceptionally complex and difficult to define narrowly.[8] One potential definition is given by Johan Galtung, who states "*that violence is present when human beings are being influenced so that their actual somatic and mental realizations are below their potential realization.*"[9] The Latin root words for our English word *violence* come from the combination of *vis* (force) and *latus* (to carry).

[8] Vittorio Bufacchi, *Violence: A Philosophical Anthology*, eds. Vittorio Bufacchi (Basingstoke, Hampshire: Palgrave Macmillan, 2009); David Riches, *The Anthropology of Violence*, eds. David Riches (Oxford: Basil Blackwell, 1986). Cf. Randall Collins, *Violence: A Micro-Sociological Theory* (Princeton: Princeton University Press, 2009); Bruce B. Lawrence and Aisha Karim, eds., *On Violence: A Reader* (Durham: Duke University Press, 2007).

[9] Johan Gultung, "Violence, Peace, and Public Research," *Journal of Peace Research* 6, no. 3 (1969): 168. Emphasis original.

Etymologically, it makes sense to follow the definition above because it literally means to carry force. While this may be an acceptable definition of the physical action, it does nothing in explaining the metaphysics and psychology behind these forces. Human violence, and therefore technological violence, is much more than mere force. My concern with violence (and its implements) has more to do with the justified *violation* of sacred space, the body, and mind than it does with physics and PET scans.

As we will see in the chapters ahead, violence is so much more than irrational aggression, but in various ways an outflow of devotion and ideology. In a phenomenological sense, *violence* is very close to *strength.* When the human's natural strength is not *powerful* enough to *force* the desires of one human upon another, it requires an implement. Readers of Hannah Arendt will have already picked up on the careful distinctions within the italicized words above. Arendt is careful to show her reader that *violence* is either about the multiplication of natural strength or a substitute for such.[10] Arendt also notes, violent actions arise for a host of psychological and physiological reasons, and not all of which are unjustified. According to Arendt, sometimes violence is the only way to "set the scales of justice right again."[11] While not an objective truth, there is evidence for what she claims, for example, the process of decolonization can be quite violent (cf. Fanon). Already one can see the great complexity and controversy at hand. Thus, I suggest that violence and its implements exist on a gradient. The question of justified violence therefore must always be under analysis and never prima facie acceptable as justified. In fact, this is the legal and ethical reality of our world. Even in cases of self-defense there exists a chain of accepted principles (innocence, imminence, avoidance, proportionality, and reasonableness) that, if broken, forfeit the defendant's *right* to self-defense. Of course, this varies from state to state and constitution to constitution. The point is we must always evaluate violence within a situated context and upon a gradient–it should never be *always justified.* This is where technology comes into play, quite literally as a mediator between two (or many more) moral agents.

---

[10] Hannah Arendt, *On Violence* (New York: Harcourt Inc, 1970), 46.
[11] Arendt, *On Violence*, 64.

Technology outflows from the desires of humans to defy limitation, and force one will upon another, whether the object is human are not. But the reader must hear the sirens echoing throughout the dark hallways of history–causation does not equal justification. Just because violence is necessary at times (more to come on this later), does not make it pristine.

When Frantz Fanon writes in *The Wretched of the Earth* about decolonization, he labels such as a process of two opposing forces. He writes, "The naked truth of decolonization evokes for us the searing bullets and bloodstained knives which emanate from it. For if the last shall be first, this will only come to pass after a murderous and decisive struggle between two protagonists."[12] These words have a devotional nature to them. What I mean by that is this, there is a *belief* that enframes them and allows them to not only formulate words of violence on a page, but to enact them on other human beings either psychologically or physically. Violence and its implements in technologies must have a belief system that functions as a psychological lever to propel thoughts into actions. Indeed, but how are we to conceive violence ontologically?

Newton Garver, in his essay "What Violence Is",[13] frames some helpful modalities for our present discussion. He gives four typological forms for understanding violence. To help visualize Garver's typology, I have put them into a matrix (see Fig 1).

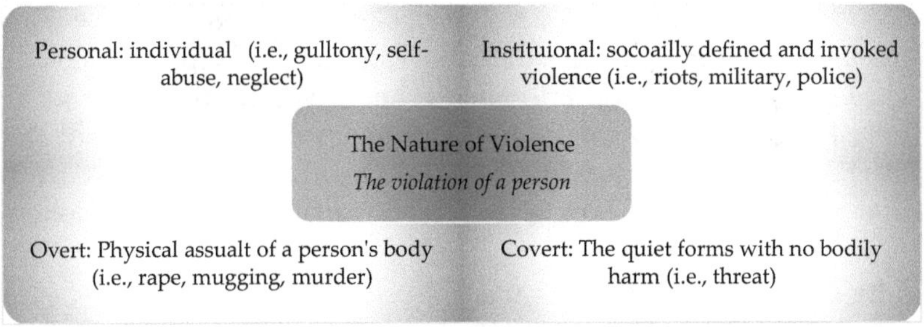

Fig. 1. Garver's Typology of Violence

[12] Frantz Fanon, *The Wretched of the Earth*, trans. Constance Farrington (New York: Grove Press, Inc, 1963), 37.

[13] Newton Garver, "What Violence Is," in *Violence: A philosophical Anthology*, ed. Vittorio Bufacchi (Basingstoke, Hampshire: Palgrave Macmillan, 2009), 170–182.

The forms of violence that Garver outlines show the philosophical complexities when trying to discern the nature of violence. In this framing, violence does not always equal force, nor are they as phenomena equal as mentioned above. These forms of violence are not static either. They cross lines fluidly, and the reality of technologies like death machines can encompass all four modalities. But I still think there is more to be said about the nature of violence, for it is not only the physical or psychological violation of a person, but a *mechanism of justification for the violation of another person.* This nuance allows for the sociologist (why is there violence), the psychiatrists (what causes violence), and the philosopher (what is violence) to dialogue about these concerns while understanding that violence is a term of moral connotation. I bring all these questions to the table in examining the modern state of violent technology. All of which are in tension with one another, and they should be. As a Christian, I believe this tension exists to ensure that justice is not at the expense of utility or design. For as we will see in the following chapters, implements exist to ensure that the killing is asymmetrical. However, this is not how such violence is framed in much of the media surrounding ethics, security, and warfare in the West.

As noted in Garver's matrix, we deem some forms of violence "necessary" at times (i.e., institutional). It is the job of citizens to discern to what degree violence is justifiable and give a clear evaluation of such acts. Such a task requires scales to weigh each case. Not to make things more esoteric than they already are around the philosophical studies of violence, but once we integrate technology into the analysis, the modalities become even more complicated. The psychological mechanisms for the justification for violent ends and the religious nature of the call for violence through implements all create a cycle of violence.

Christian Views of Violence

As we will see in the examination of just war theory, there is no historical consensus as to what constitutes legitimate forms and actions of violence. Based on the historical context and how that influences the exegesis of passages like Romans 13, perspectives vary from generation to generation. While the particulars of such exegesis do not concern us here, it is helpful to understand how perspectives of war and violence have adapted over the centuries of Christian thinking. For violence, and its most structured form in warfare, compromise has always been in play.

Christian views of violence through the centuries have swung from nonviolence to revolutionary violence. The compromise comes into tension with the Christian's reading of passages that deal with submission to authority. The political Augustinianism of the Middle Ages saw the government as God ordained and thus the state legitimately, under the hand of God, could use the sword both in a domestic and foreign context. Political violence became a ministry of the Church. Following the Roman thinking about states and violence, there were three compartments for violence in Christian traditional thought: (1) state violence was not equal to individual violence, (2) violence and force are not the same, and (3) the state's use of violence is not necessarily just or unjust. Per this last point, many theologians, continuing today, hold a view that the spiritual dimension to state sanctioned violence lends it to being pristine. What is commanded by God and executed by the state is just, and not necessarily the business of the Church (i.e., John Calvin).[14] The assumption that holds from this logic is that the state is a priori possible of using deadly force and civil power and doing so justly. Here we see the continuing tension to clear the charges against the state and conscious of the ecclesiastical body. The primary concern for allowing state sanctioned violence is that by tying the hands of the state, in the name of Christian pacifism, dooms it to be overtaken. Canon 3 of the Council of Arles (c. 314) states, that "Concerning those who lay down their weapons in peacetime, [*in pace*][15] be it resolved that they be excluded from fellowship."[16] Christians during this new relationship of church and state where not to disobey orders to serve in the military, for it was their duty to obey the state which was granted its authority under God.

Augustine communicated this view (cf. Ambrose of Milan c. 339–387),[17] in response to the sack of Rome by the Goths in 410. His tome *The City of God* was the first Christian history to address the tension between the dilemma of human sin nature and the life of the believer

[14] Cf. John Calvin, *Institutes of the Christian Religion*, ed. John T. McNeill, trans. Ford Lewis Battles, vol. 1, The Library of Christian Classics (Louisville, KY: Westminster John Knox Press, 2011), 1215.

[15] Most likely refers to the peace between the Church and Empire.

[16] See commentary on this canon by Adolf von Harnack in, *Militia Christi die christliche Religion und der Soldatenstand in den ersten drei Jahrhunderten* [Milita Christi: The Christian Religion and the Soldier-hood in the Frist Three Centuries] (Tübingen: Mohr [Paul Siebeck], 1905), 87–88.

[17] Augustine, "Letter 51," trans. Henry De Romestin, in *The Nicene and Post-Nicene Fathers*, 10.

inside society. In response to count Boniface, Augustine wrote, "Do not think it is impossible for anyone to please God while engaged in active military service."[18] Not only did Augustine see military service as theologically acceptable, but war was necessary considering the evil forces present within this world. Commenting on the beatitudes: "Blessed are the peacemakers; for they shall be called the children of God," he writes,

> Therefore, even in waging war, cherish the spirit of the peacemaker, that, by conquering those whom you attack you may lead them back to the advantages of peace for our Lord ... As violence is used towards him who rebels and resists, so mercy is due to the vanquished or the captive, especially in the case in which future troubling of the peace is not to be feared.[19]

A casuistry view of violence and warfare is also seen in the logic of Thomas Aquinas. Aquinas' seven conditions for just war are no less problematic. Much like Augustine, and others before him, the entirety of the argument rests on assumptions about biblical justice, or in Aquinas Aristotelian justice. War as a means for preventing further evil, or even as reconciliation, was a legitimate force, because the state is a servant of God.

Another view is that of nonviolence, which goes back to the pacifism of the early followers of Jesus. Until the fourth century, this was the official position of the early church.[20] Athenagoras, an Athenian philosopher, and Tertullian, a jurist, argued that an aversion to violence and exemption to military service is a legitimate path for Christians. Tertullian reflects on this tension and conflict many early Christians had between loyalties to Christ and flag,

> And shall the son of peace take part in the battle when it does not become him even to sue at law? And shall he keep guard before the temples which he has renounced? Shall he carry a flag, too, hostile to Christ? Then how many other offenses there are involved in the performances of camp

---

18 Augustine, "Letter 189," trans. J. G. Cunningham, in *The Nicene and Post-Nicene Fathers*, 1.
19 Augustine, "Letter 189," cf. *The City of God*, bk 19, 12–27.
20 J. M. Hornus, *Évangile et Labarum* (Geneva: Labor & Fides, 1960).

> offices, which we must hold to involve a transgression of God's law, you may see by a slight survey.[21]

For Tertullian, it is not simply about the physical acts of violence or the use of violent implements on other image bearers, but the whole edifice which calls into question the ethics of the Christian. Are they serving Christ or the crown? They met the stand of nonviolence in the wake of growing armies in the third century with great violence, leaving many Christians martyred. Despite this military martyrdom, conscientious objection to military service continued throughout the history of the church. Figures like St. Martin of Tours, Francis of Assisi, the brothers of Waldo, and John of Leyden provided examples of nonviolence for the local church. The theological logic that undergirds this movement is that there is an imperative in Christological ethics to let love be the force that overcomes evil. In this view, violence is by nature contrary to love. What this means is that by resorting to violence, one forgoes faith that God can change the heart of the other. This theological view of violence lends itself to two ethical approaches. The first approach wrestles with the moral voice and agency of the nonviolent individual. When the desire and devotion of the individual's inner voice is one of peace, it will lead to external actions of peace. The second deals with institutions of violence (i.e., government and military). When institutions are stripped of their implements, enforcing their will and power are at risk. Stripping the police and armies of their weapons of violence (through nonviolent means) will ultimately lead to the dismantling of violent force and injustice. How do we test this theory? One way is to look at advocates for theological nonviolence like Martin Luther King Jr.

Martin Luther King Jr is famously known for being an advocate for nonviolence, but what were the objects of his platform? He warned about the "Giant Triplets": militarism, materialism, and racism–these were the core of his message. Here we have a clear message about the *who* and *what* of nonviolence. While in Birmingham jail in 1963, King wrote to a fellow minister about the *demonstration* in Birmingham. In his letter, he wrote, "You deplore the demonstrations taking place in Birmingham. But your statement, I am sorry to say, fails to express a

[21] Tertullian, "The Chaplet, or De Corona," 11, In *Latin Christianity: Its Founder, Tertullian*, eds. Alexander Roberts, James Donaldson, and A. Cleveland Coxe, vol. 3, The Ante-Nicene Fathers (Buffalo, NY: Christian Literature Company, 1885), 99.

similar concern or the conditions that brought about the demonstrations…It is unfortunate that demonstrations are taking place in Birmingham, but it is even more unfortunate that the city's white power structure left the Negro community with no alternative."[22] While King was no innovator of nonviolent protest,[23] he is perhaps the most famous in modern memory for amassing such for theological and political reasons. The political platform of nonviolence, for King and others, was to draw out the spiritual injustice of inequality, and in King's historical context the unbiblical nature of racism. Instead of implementing violent resistance, King used the life of the mind to force the thinking of society to reconsider its ways. As Walter Benjamin reminds us in "Critique of Violence," "a totally nonviolent resolution of conflicts can never lead to a legal contract."[24] It is through language and its translation and understanding that conflicts are resolved, or at least until the next conflict arises. Benjamin and King realized that nonviolence is a type of *technique* that is possibly ungovernable. Liberation goes hand in hand with reconciliation and this is the ultimate *telos* of what historical figures like Jesus, Gandhi, King, and the Fellowship of Reconciliation (FOR)[25] sought to bring about.[26]

From a theological rationale, "resist not the evildoer," is appealing and desirable. But the biggest challenge to its movement is what we know about the internal and destructive nature of humans. The Bible does not appease modern sensibilities of *goodness* within human nature; even John records that "But Jesus, on His part, was not entrusting Himself to them, for He knew all men" (John 2:24). Humans are inconsistent. In one vein,

---

[22] Martin Luther King Jr., "Letter from Birmingham Jail," 128–30.

[23] See Faith Hoolsaert, Martha Prescod Norman Noonan, Judy Richardson, Betty Garman Robinson, Jean Smith Young, and Dorothy M. Zellner, eds. *Hand on the Freedom Plow: Personal Accounts by Women in SNCC* (Chicago, IL: University of Illinois Press, 2012); Andrew Young, *An Easy Burden: The Civil Rights Movement and the Transformation of America* (Waco, TX: Baylor University Press, 2008); J. Deotis Roberts, *Liberation and Reconciliation: A Black Theology*, 2nd ed. (Louisville, KY: Westminster John Knox Press; 2005); and Maegan Parker Brooks and Davis W. Houck, eds., *Speeches of Fannie Lou Hamer: To Tell it Like it Is* (Jackson, MS: University Press of Mississippi, 2011).

[24] Walter Benjamin, "Critique of Violence," In *Walter Benjamin Selected Readings, Volume 1: 1913–1926*, eds. Marcus Bullock and Michael W. Jennings (Cambridge, MA: Harvard University Press, 2004), 243.

[25] See *The Fellowship of Reconciliation*, https://forusa.org/forhistory/.

[26] See Walter Wink, *Jesus and Nonviolence: A Third Way* (Minneapolis: Fortress Press, 2011), 1–9.

we can write massively influential books about nonviolence, take John Howard Yoder's *the Politics of Jesus* for example, and in another vein the same man can sexually abuse 13 women.[27] How can both be realities of the same man? This brutal reality does not make the theology of Yoder's work untrue, but it does show the dissonance between belief and practice. For so long women have been both the object of liberation and the victim of those who claim to set them free. As Judith Butler reminds us, "the struggle against violence accepts that violence is one's own possibility."[28] I think womanist theology understands this better than theologies of liberation portrayed by the men above. For it is their historical and present reality to be the victim and survivor.

There is no more sobering realism of violence and its implements than the experience of those that live it. After all is this not the way of Jesus, to sit with the broken, abused, marginalized (often women and children in the first century), and listen to their story. To put my concern in the phrasing of Levinas, it is to resist the "temptation" to neglect the "face" of the other.[29] As much as I would like to champion perspectives of nonviolence, I worry that doing so ignores the voice of those men, women, and children who live in these conflicts, and deal with the moral injury long after the abuse has passed. Nonviolence is not the absence of conflict. Nor should we conflate nonviolence with passivity, for that is not the cause. To avoid such confusion, I prefer to use the language of *antiviolence* (active resistance)[30] as found in the work of David C. Cramer and Myles Werntz. While Cramer and Werntz are specifically focused on sexual violence, the concept could also be used in a broader political discussion about violence. Jesus' resistance of violence was not about passivity, but about passion (i.e., suffering). If what we mean by

[27] Michelle Sokol and The Elkhart Truth, "Mennonite Seminary Apologizes to Victims of Famed Theologian John Howard Yoder," *National Catholic Reporter*, April 9, 2015, https://www.ncronline.org/news/accountability/mennonite-seminary-apologizes-victims-famed-theologian-john-howard-yoder (accessed Feb 22, 2023).

[28] Judith Butler, *The Frames of War: When Is Life Grievalbe?* (New York: Verso Press, 2009), 171.

[29] I am speaking loosely here of the philosophy communicated in Emmanuel Levinas, "Peace and Proximity," in *Basic Philosophical Writings*, eds. Adriaan T. Peperzak, Simon Critchley, and Robert Bernasconi (Bloomington, WA: Indiana University Press, 1966), 161–69.

[30] David C. Cramer and Myles Werntz, *A Field Guide to Christian Nonviolence: Key Thinkers, Activist, and Movements for the Gospel of Peace* (Grand Rapids, MI: Baker Academic, 2022), 129.

nonviolence is simply submission and subjection, then it cannot be Christian, nor should it be an ethical option for society.

I want to pause here, acknowledging the difficulty of the subject at hand, and clarify what could be misunderstood in Christian visions of violence, atonement theory, and suffering. I am a theologian. I believe God can reconcile us to himself and use what men meant for evil for good (Gen 50:18). But there is so much violence that can (but should not) be justified in the name of this theology. For it can make something like the brutal and bloody crucifixion of Jesus, a model of valor. I agree that the death of Jesus was for my good, but the exegetical key to understanding the theory of atonement is that his death is volitional and not by *force*. Nothing I do, or any other Christian may do, produces the same salvific results as the atonement of Jesus on the Cross. Therefore, knowing this, how can one ever say causing, allowing, or taking part in the involuntary suffering of another is 'for the greater good.' To put it in Pauline language, "may it never be."

The Process of Violence

Violence, much like technology, must be seen as a gradient and not as black and white. It is all too common in folk conversation and theory to say things like, "I do not condone violence" or "I am nonviolent." But what does that even mean? Certainly, there are times when everyone is for violence (cf. Arendt and Fanon). For example, when a surgeon forcibly enters your body to remove a cancerous mass, we are generally for this act of violence, even though it will cause pain and short-term suffering. Granted that this is more so force than it is violence proper. However, there are also violent acts upon the gradient we are less comfortable with or even flat-out reject, and rightly so. Rape, for example, is not only violence against the biological body but also the mind. The force of power lingers long after the physical damage, often leading to a recursive cycle (or step) toward further violence.

In this framework, violence and technology are linked in a process. The creation of a tool does not happen *ex nihilo*, it comes from the enframing of the creator to achieve a desired outcome. Therefore, we ought to see violence and technology on a gradient and entangled. What comes forth as a technological implement is first begotten in the cycle or desire to overcome some limitation. Thus, it is difficult, if not impossible, to assess a piece of technology in isolation from its rationale for creation.

More to this, the desired creation alone does not guarantee its usage or application. Gradients, such as the one above, rightfully make people uncomfortable.

In a strict sense, this book is about violent technologies; thus, it covers in depth the nature and critique of AI and robots within war theaters. Yet, to make such assessments, one must go back to the fundamental rationale for creating such technology. I will cover both the process and reasons why society makes room for violent technology and how it laterally impacts both creator and creation. I will also give a theological and philosophical critique of the nature of violence and its foundation in religious thought.

What is Technology?

Heidegger, universally hated as he may be, gets closer to the essence (*wesen*) of what technology is in its being and its extension. The essence of a technology relates to how it endures through presence.[31] Heidegger's view of the essence of technology balances McLuhan's extension of man theory–that is, a technology is primarily instrumental. Technology is no mere means to an end, but the logical end of a process or cycle that begins with enframing and desire. Technology's essence is in what reveals (*aletheia*). One of the missteps in current scientific thinking goes back to the distinction Aristotle makes in *Nicomachean Ethics* (bk. VI, ch. 3–4), between *episteme* and *techne* as means of truth. Mortimer Adler points this out in his work, *The Conditions of Philosophy*, the conflation of these terms is a small hint as to why modernity misunderstands the meaning and revealing of technology.[32] For Heidegger, technology is a mode of revealing the truth. The truth about what?

Here is where Heidegger is less helpful; what truth is revealed is primarily related to the "self" or "the man." He writes, "In Enframing, that concealment comes to pass in conformity with which the modern technology reveals the real standing-reserve. This work is therefore neither only a human activity nor a mere means within such activity… And it cannot be rounded out by being referred back to some

[31] Martin Heidegger, *An Introduction to Metaphysics*, trans. Ralph Manheim (New York: Double Day, 1961), 59.

[32] Mortimer Adler, *The Condition of Philosophy: Its Checkered Past, Its Present Disorder, and Its Future Promise* (New York: Dell Publishing Co., 1967).

metaphysical or religious explanation that undergirds it."[33] Now that does not mean that the actions are based on a warranted and justified belief from an epistemological or empirical stance, but rather we can (in some regard) begin to understand why one human can look into the eye of another and dehumanize them, even on an extreme scale as in the horrific acts of genocide.

Technology as Religion

Egbert Schuurman, professor of reformational philosophy at the Technological Universities of Delft and Eindhoven in the Netherlands, is of great help for readers seeking to understand enframing. As mentioned above, technology is not devoid of devotion, but, in fact, is often mired in religious thought. I bring in Schuurman here because a majority of his work is dedicated to the critique of this very phenomena. In his 2003 work, *Faith and Hope in Technology*,[34] Schuurman traces the contours of this movement that arises in the absence of traditional Christian metaphysics. He labels this "technicism" which emerges as an attempt to *force* one deity for another. In this process technology becomes a method of control to shape reality in the image of man. More on this to follow below, but first let us consider a thesis from Erik Davis' work *TechGnosis*.[35] Davis' work emerged in the 1990s, but don't let the dated tech references fool the reader into believing that the work is irrelevant for analyzing the theology and philosophy of technology in the present. The vision of a future utopia is not primarily a Western vision. The primary vision of technology is eschatological, whether creators and tech billionaires claim a religion or not, "this secular image was framed all along by Christian myths: the biblical call to conquer nature, the Protestant work ethic, and in particular, the millennialist vision of a New Jerusalem…"[36] The irony here is that many Christian philosophers and theologians, such as Jacques Ellul, are overwhelmingly pessimistic about technology, but technology is a religious and overtly Christian vision. For

---

[33] Heidegger, "The Question of Technology," 21.

[34] Egbert Schuurman, *Faith and Hope in Technology*, trans. John Vriend (Toronto: Clements Publishing, 2003).

[35] Erik Davis, *TechGnosis: Myth, Magic & Mysticism in the Age of Information* (Berkley, CA: North Atlantic Books, 2015).

[36] Davis, *TechGnosis*, xix.

Ellul, in his critique of *technique*, the primary concern is that technological implements have taken a life of their own.[37] Similar to what Schuurman labels as *technicism*, such a view sees technology as the extension of power and economic means. Ellul, and many contemporary Christian thinkers agree that the embracing of the technological is the subverting of the Christian calling to "subdue and rule." Of course, this reading is a very confused and misinformed reading of the cultural mandate. The creation of new technologies is perhaps the most ancient practice of religion; with new instruments come new gods. As we will see in later chapters, not all these gods are benevolent. Belief drives creation and there is no creation that comes forth free of myth.[38]

Christian views of the eschaton have driven overtly utopian visions and dreams about the symbiosis of technology and theology. The marrying of apocalyptic visions and new media is not always a match made in heaven, in fact, the history of theology and tech is one of justified violence. Davis claims that writing is a machine. In that case, the Bible is both a technology of life and death. Davis is correct here when he writes, "Fundamentalist certitude to the contrary, working with Scriptures is a tricky and open-ended process, because the machinery can never contain and control all its own meanings."[39] The history of biblical interpretation clearly justifies this claim, as well does the history of translating the Bible into English. Fredrich Farrar's lectures on the history of interpretation (c. 1885) capture these temptations clearly in his analysis. He writes, "Few are the translators, fewer still the exegetes, who have been so free from various idols of the cave, the forum, and the theatre as to abstain from finding in the Bible thoughts which it does not contain and rejecting or unjustly modifying the thoughts which indeed are there."[40] For this reason, we ought to pause briefly to examine the Christian doctrine of original sin and how that brings to bear on technology and its implementation in enacting the desires of interlocutors.

---

[37] Jacques Ellul, *The Technological Society* (New York: Knopf, 1964).

[38] Or as Michel Serres has written, "The only pure myth is the idea of a science devoid of all myth," cited in Bruno Latour, *We Have Never Been Modern*, trans. Catherine Porter (Cambridge, MA: Harvard Press, 1993), 93.

[39] Davis, *TechGnosis*, 26.

[40] Frederic William Farrar, *History of Interpretation* (London: Macmillan and Co., 1886), 5.

An Overview of Sin Nature

In Christian complementary theology there are four major categories that frame how the narrative of Scripture are read and thus how doctrines are derived. The first is creation. God is the source of reality and the author of all underlining metaphysics and physics. Therefore, the Hebrew Bible begins with an epistemological claim about God as sovereign over cosmology. In the Biblia Hebraica it reads,

בְּרֵאשִׁית בָּרָא אֱלֹהִים אֵת הַשָּׁמַיִם וְאֵת הָאָרֶץ

At the beginning God created the heavens and the earth.[41]

The axiological assumptions in this one sentence enframe the rest of the reader's theology. Unlike methodological naturalism, which pervades most of current scientific research, God is considered a formal cause of life and meaning. Much like the newborn forced into the world through the birth canal or through surgical extraction, humanity does not have a choice in the worlds shaping and its guiding tenants. In Christian cosmology, God is the creator and sustainer of this world. Yet, something is awry. In fact, shortly after the creation narrative humans are forced out of the garden in which they were placed.

Colloquially, this episode in Genesis 3 is known as the "fall" narrative. As we move into a theological critique of violence, it is imperative that we balance God's decree within Scripture with a proper theology of human nature. In Genesis 2–3, we see a theological narrative[42] about how humans deal with autonomy and the quest for knowledge, even at the detriment to their physical and spiritual wellbeing. In the history of interpreting this text, many theologians are tempted to do great violence to it by overloading the narrative with the language of sin. Certainly, primal sin is a part of the theology at play, but closer to the heart of the text is desire.

---

[41] Author's translation.

[42] This language is not a commentary on the historical Adam and Eve. I believe they were, in fact, historical people. For reconciliation with modern genealogical science, see S. Joshua Swamidass, *The Genealogical Adam and Eve: The Surprising Science of Universal Ancestry* (Downers Grove, IL: IVP Academic, 2019.

וַתֵּרֶא הָאִשָּׁה כִּי טוֹב הָעֵץ לְמַאֲכָל וְכִי תַאֲוָה־הוּא לָעֵינַיִם וְנֶחְמָד הָעֵץ
לְהַשְׂכִּיל וַתִּקַּח מִפִּרְיוֹ וַתֹּאכַל וַתִּתֵּן גַּם־לְאִישָׁהּ עִמָּהּ וַיֹּאכַל׃

When the woman saw the tree was **desirable** for food, it was a **delight** to the eyes and **desirable to make one wise;** she took its fruit and ate, and she gave also to her husband who was with her, and he ate.[43]

The brokenness of human nature does not begin with maliciousness, but with innocence. Adam and Eve did not know the horrors that awaited them outside of the garden. They perceived, in ignorance, that the garden was a prison, and the tree of knowledge was their enlightenment. They did not realize their perfect shalom with God, creation, and one another[44] until it is lost through action upon temptation. As Dietrich Bonhoeffer has reflected, this conversation in Gen 3 is "the first conversation about God, the first religious, theological conversation, it is not prayer or calling upon God together but speaking about God, going beyond him."[45] The couple could have gone to God and asked Him why, but that is not in the narrative because revolt is within the anatomy of their heart.

It roots the original temptation in a desire to know (*yada'*). As in the Epic of Gilgamesh, to "know" has a sexual undertone to it. It is an experimental knowledge. This is not a commentary on the desire of sexuality in general, least the reader thinks that God is opposed to such. Rather, it is about the temptation to go outside of God's knowledge and revelation. Thomas McCall writes, "The awful irony of temptation and sin is that so often we begin to distrust God's word and doubt the sincerity of goodness of his intentions for us *at exactly the point where he is trying to protect us*."[46] East of Eden is where violence begins. Humans are forced out of the garden through the power of God. Animals are killed

---

[43] Author's translation.

[44] See John Goldingay, *Old Testament Theology: Volume 1: Israel's Gospel* (Downers Grove, IL: IVP Press, 2003), 136; Walter Eichrodt, *Theology of the Old Testament*, trans. J. A. Baker, vol. 1 (Philadelphia, PA: The Westminster Press, 1961), 457; Eugene H. Merrill, *Everlasting Dominion: A Theology of the Old Testament* (Nashville, TN: Broadman & Holman, 2006), 203.

[45] Dietrich Bonhoeffer, *Creation and Fall, Temptation: Two Biblical Studies* (New York: Simon & Schuster, 1997), 76.

[46] Thomas McCall, *Against God and Nature: The Doctrine of Sin*, Foundations of Evangelical Theology (Wheaton, IL: Crossway, 2019), 122. Emphasis original.

for the first time. Shortly after, Cain murders his brother Abel in a jealous and misguided desire to fulfill his inner spirit. Indeed, but it is through violence that God will redeem his children and fulfill the proto gospel in Genesis 3:15: "I will put enmity between you and the woman, and between your offspring and her offspring; he shall bruise your head, and you shall bruise his heel."

Does that entail that God is the author of violence? I think not. This again goes back to theories of causality, yes metaphysics once again. In theological circles this is often referred to as "divine authorship" or "causation of sin." F. D. E. Schleiermacher, in his theological formulation of *Gefuhl* "feeling of absolute dependence" he sees a dual nature to the cause of sin. It is partly within the desire of the human and "partly outside our own being."[47] In that sense, according to Schleiermacher, God is the formal cause of sin. Sin nature is intrinsic to the human and original to all of us. this is important to him because of his views of supralapsarianism–God must hold divine omnipotence.[48] Even though Schleiermacher's rationale for his view is to uphold a high view of the power of God, it is not derived from the text itself, in fact, it is quite a violent reading of the text, forcing a particular theological lens, in this case a Calvinistic one. More to this, it violates the principle of proportionate causality (PPC) which states, "Effects must be proportionate to their causes and principles."[49] Since there is no darkness in, God (or sin), He cannot cause it to be. In order for God to cause or create sin it must be in Him formally, virtually, or eminently. The rejection of this view has been present in Church history long before it was put into ink. The Synod of Orange (529), Bonaventure, and Aquinas all reject the cosmic duality of original sin.[50] God is neither the direct nor indirect cause of sin, yet nothing happens without His appointment.[51] The destructive nature of sin (brokenness and failure) is not the design

[47] F. D. E. Schleiermacher, *The Christian Faith* (Edinburgh: T&T Clark, 1989), 271.

[48] Schleiermacher, *The Christian Faith*, 335.

[49] Thomas Aquinas, *Summa Theologica*, trans. Fathers of the English Dominican Province (London: Burns Oates & Washburne, n.d.).

[50] Henry Bettenson, ed., *Documents of the Christian Church*, 2nd ed. (Oxford: Oxford University Press, 1963), 62; Bonaventure, "Breviolquium," bk. III. Line 1, in *The Works of Bonaventure*, trans. Jose de Vinck, vol. 2 (New York: St. Anthony Guild Press, 1963), 110–121; Aquinas, *Summa Theologica*, IaIIae, Q79. 1.

[51] Cf. G. C. Berkouwer, *Studies in Dogmatics: Sin* (Grand Rapids, MI: Wm. B. Eerdmans-Lighting, 1971), 27–31.

of God. However, violence naturally follows desire and devotion. It is difficult to read the Old Testament or the New Testament and not see the reality of this in human nature.

Technology and Rebellion

Earlier I mentioned violence and technology are intertwined. That statement requires further theological unpacking. In the section above, I have painted a broad stroke of the doctrine of "original sin," and it needs to be challenged in light of the teleological ends of violent technologies and, thus why I sense many Christians and theologians have made peace with violence and its implements, especially in light of mechanized warfare under views of Just War Theory (JWT).

The traditional theological framing of original sin comes from a few prominent theologians: Augustine, Paul Tillich, and Reinhold Niebuhr. In essence, all see *sin* as human rebellion against God. These views have major problems for the task at hand. For Augustine, the transmission and predisposition of sin stems from the angelic revolt of Satan. Rebellion through self-transcendence is every human's story. Augustine, utilizing the biological knowledge of his era, communicates this in his massive work *The City of God.* Through the sexually transmitted seed of Adam, all humans inherit a sin nature (Cf. *Against Julian*).[52] This reading became the primary heuristic for Christian theology and is perhaps still the primary reading today through the protestant admiration for writers like John Owens, John Edwards, and John Calvin. Sensuality, in a physical sense, is the primary driver of sin nature according to the Augustinian view.[53] The logical trajectory of this view allows for unacceptable actions such as war rape to result in a *natural* tendency to allow the physical nature of the human to dominate over the higher (i.e., spiritual). Humans are simply corrupted and acting in occurrence with this corruption. This view becomes even more problematic (per evidence

---

[52] Augustine of Hippo, "The City of God," in *St. Augustin's City of God and Christian Doctrine*, ed. Philip Schaff, trans. Marcus Dods, vol. 2, A Select Library of the Nicene and Post-Nicene Fathers of the Christian Church (Buffalo, NY: Christian Literature Company, 1887).

[53] Augustine of Hippo, "Letters of St. Augustine," in *The Confessions and Letters of St. Augustin with a Sketch of His Life and Work*, ed. Philip Schaff, trans. J. G. Cunningham, vol. 1, A Select Library of the Nicene and Post-Nicene Fathers of the Christian Church (Buffalo, NY: Christian Literature Company, 1886), 55–59.

in the next chapter) when theologians hold this view and claim that because Scripture accommodates this nature in the context of ancient war practice, it is now a pristine action. To that, I respond with an emphatic *no*![54]

Another troublesome view of original sin is Reinhold Niebuhr's anxiety view presented in *The Nature and Destiny of Man*.[55] Niebuhr channel's the Augustinian paradigm through a psychological lens. The source of sin for Niebuhr lies within what he calls "anxiety" the *natural* state of humans which is in flux between limitedness and freedom. This state is not *sinful* but rather stands as an Erlenmeyer flask and when worry or anxiety is added into the mix, sin results. A lack of security is the primary contention for Niebuhr (cf. Kierkegaard).[56]

This logic is present throughout Christian thought, and justifiably so, but this framing, again, is not helpful. To see every sin as a rebellion against God (i.e., primal sin), is too abstract and at the same time too simplistic. Here I want to introduce a heuristic that is helpful in seeing the nuance of violence in the Bible and, hopefully in discerning a theology of violent technologies.

Judith Plaskow, Valerie Saiving, and Susan Nelson's feminist critique of Niebuhr reverse the development of the nature of sin.[57] Instead of seeing sin as a rebellion against God, the feminist sees the nature of sin as a rebellion against creation and thus a rebellion against God. How is this different? One of the major weaknesses of Niebuhr's theory is that it is usually only applied to those in positions of power and luxury–those that have the means to enact their pride or self-transcendence. It is a failure to see the lateral effects of ill-being. All is absorbed into "spirit," including nature. All of which is grounded or rooted in God's will. Only those in positions of power are primarily concerned with defining such.

---

[54] There are more theological problems with this view, but I am limiting my critique of such for now. For more critique see Millard Erickson, *Christian Theology*, 3rd ed. (Grand Rapids, MI: Baker Books, 20013), 515–30.

[55] Reinhold Niebuhr, *The Nature and Destiny of Man* (New York: Scribner, 1941), 1:182.

[56] Søren Kierkegaard, *The Concept of Anxiety: A Simple Psychologically Orienting Deliberation on the Dogmatic Issue of Hereditary Sin*, ed. and trans. Reider Thomte (Princeton, NJ: Princeton University Press, 1980).

[57] Valerie Saiving Goldstein, "The Human Situation: A Feminine View," *Journal of Religion* 40 (1960): 100–112; Judith Plaskow, *Sex, Sin and Grace: Women's Experience and the Theologies of Reinhold Niebuhr and Paul Tillich* (New York: University Press of America, 1980); Susan Nelson Dunfee, "The Sin of Hiding," *Soundings* LXV, no. 3 (1982): 316-27.

Those in marginalized positions, historically women, are concerned with defying limitation. The fundamental sin is against the good creation that God has made and thus entails a sin against God. This is an important distinction because it does not abstract ethical actions upon moral agents. There must be an account for freedom, guilt, and responsibility with the relation to creation and the creator. The violence of rape, war, economic oppression, and so on cannot be abstracted, or explained away as the mere teleological ends of the desire to transcend. Their blood calls for accountability.

Jacques Ellul, in his order of necessity, notes that violence, whether justified of rationalized is all equal. Violence only leads to more violence. Regrettably, Ellul never defines violence.[58] In a similar vision, theology's that make concessions for violence (i.e., *it is* in our nature) always make room for the justification of sin, oppression, and marginalization of the *other*. A theology of violence, technology, and violent technology cannot accept the destruction of another entities well-being simply because there exist *natural* desires for such. If even one were to accept the original sin thesis of Augustine or Niebuhr, what good comes from it for a theological view of justice, which is the antithesis of violence?

Before Augustine, there was Irenaeus, and he too wrote about the perplexed nature of reconciling a good God with a fallen world. Irenaeus posited that the nature of humanity is that of an infant. Humans, in this theory, are in the process of emergence, from being biologically made to being made in His image. The potentiality is there from the beginning, but that does not mean it fated the outcome to be so. This emergence framework provides a foundation for a different perspective, and one that allows for complexity in dealing with the problem of evil and closely fits into modern behavioral psychological and evolutionary theories about violence.

Irenaeus Eibl-Eibesfeldt (not the same Irenaeus) developed a theory that reflects the logic above in light of human and animal behavior.[59] Eibl-Eibesfeldt's research, further substantiated by the scientific work of Adrian Raine, claims that humans are biologically structured toward aggression. Physical violence is the release of aggressive impulses, as either a means of pleasure or survival. There is another biological drive

---

[58] We will return to Ellul in the next chapter.

[59] Irenaeus Eibl-Eibesfeldt, *Love and Hate* (New York: Holt, Rinehart and Winston, 1971).

within humans and animals, that of bonding. The reality of sin, regardless of source, is that humans live in the tension between love and hate, violence and forgiveness, and control and submission. It was Eibl-Eibesfeldt's hope that humans would choose bonding over aggression. Likewise, in the person of Jesus there is hope for an alternative view of violence and its implements.

The concept of *lex talionis* "measure for measure" (i.e., physical retribution) or "eye for an eye" predates the biblical law code (cf. Ex 21:22–25; Lev 24:17; Deut 19:15–21). We find the ancient Near Eastern laws of physical retribution in the sources of Mesopotamian Code of Shulgi (c. 2112–2095 BC), Upit–Ishtar (c. 1860 BC), and Law of Hammurabi (c. 1700 BC). All of which are concerned with the limitation of escalation or retaliation.[60] Violence predates the fall and the rebellion of fallen nature. Perhaps the innateness toward aggression has more to do with creating creatures with moral agency than with pride or defiance of limitation. Whatever the reality may be, responsibility for one's actions and the resistance to revenge and violence are at the heart of Jesus' teaching on forgiveness. Violence begins and ends with the disregard for the *other's* well-being, but it is more than that, for it perpetuates a cycle of violent actions and ultimately victimization. No human person has fully experienced this more than the person of Jesus.

René Girard, Robert-Hamerton-Kelly, and James G. Williams have brought this reality to life in theories of ritual, sacrificial, and collective religious violence.[61] Jesus is the ultimate victim of violence through a

---

[60] S. West, "The Lex Talionis in the Torah" *Jewish Bible Quarterly* 21 (1993): 183–88; M. G. Kline, "Lex Talionis and the Human Fetus," *Journal of Evangelical Theological Studies* 20 (1977): 273–304; Roland de Vaux, *Ancient Israel: Its Life and Institutions*, trans. John McHugh (New York: McGraw–Hill, 1961), 149–50; Robert R. Wilson, "Enforcing the Covenant: The Mechanisms of Judicial Authority in Early Israel" in H. B. Huffmon, F. A. Spina, and A. R. W. Green, *The Quest for the Kingdom of God: Studies in Honor of George E. Mendenhall* (Winona Lake: Eisenbrauns, 1983) 59–75. For the biblical parallel's see Victor H. Matthews and Bon C. Benjamin, *Old Testament Parallels: Laws and Stories from the Ancient Near East*, rev. 3rd ed. (New York: Paulist Press, 2006), 103.

[61] Girard is the most prominent of the scholars mentioned, but that does not mean he is the only one to capitulate a theory of mimesis for collective violence. All three of these scholars remix one another's work, in a sense. All have valid points to bring to the discussion. See René Girard, *Deceit, Desire, and the Novel* (Baltimore: John Hopkins, 1976); *Violence and the Sacred* (Baltimore: The Johns Hopkins University Press, 1986); *The*

Roman remix of Persian technology (i.e., death by crucifixion). The words of Girard are worthy of serious consideration within any tome of violence and theology. One lengthy quote comes to mind.

> To recognize Christ as God is to recognize him as the only being capable of rising above the violence that had, up to that point, transcended mankind. Violence is the controlling agent in every form of mythic or cultural structure, and Christ is the only agent who is capable of escaping from these structures and freeing us from their dominance…A non-violent deity can only signal his existence to mankind by having himself driven out by violence–by demonstrating that he is not able to establish himself in the Kingdom of Violence.[62]

While I do not contend that this book is a recapitulation of Girard (or others) theory of mimesis, I do wholeheartedly agree that Jesus is the best solution to approaching the topic of violence and its implements.

Even in the word's etymology (*techne*, *logos*, and *tehcnologia*) there is the human desire to rebel against the natural. Again, that is not a moral commentary of any given piece of technology, but inherent to such creations is a desire to control, violate, or force something that without an implement might not come to be. This reality is not positive or negative. Technology can *force* crops to yield and water to purify. However, it also has the potential to *force* atoms to split. Depending on the enframing of the human interlocuter, technology either becomes a force for blessing or burden. Alongside Schuurman, David Noble in *The Religion of Technology* bears witness to the temptation for humans to use technological means for their religious ideological ends.[63] Much like our ancestors Adam and Eve we see our limitedness within a created order as a prison to escape.

Scapegoat (Sacred (Baltimore: The Johns Hopkins University Press, 1989); Robert Hamerton-Kelly, *Sacred Violence: The Hermeneutics of the Cross in the Theology of Paul* (Minneapolis, MN: Fortress Press, 1991); and James G. Williams, *The Bible, Violence, & the Sacred: Liberation from the Myth of Sanctioned Violence* (San Francisco: Harper Collins, 1989).

[62] Girard, *Things Hidden Since the Foundation of the World* (Stanford: Stanford University Press, 1987), 219.

[63] David Noble, *The Religion of Technology: The Divinity of Man and the Spirit of Invention* (New York: Alfred Knopf, 1997).

What is Violent Technology?

While humans formally caused technologies, it does not mean they are fated toward violent ends. It is the hope of this book, that by naming and clarifying the process of creating and implementing violent technologies, humans will take steps toward de-escalation of future unjustified violence. The creation of violent technologies is ultimately about the violation and *othering* of fellow humans through artificial entities. Yet, while that is the desired telos of these machines, the effect is not unilateral. The creation of violent technology necessitates that one uncreates their moral personhood. What makes one a person is not just substrates of biological material, but the reflection of ontological and non-material properties that we see (and feel) within the self. The existential dread over superintelligence and autonomous weapons systems by scholars like Nick Bostrom, Peter W. Singer, and Elke Schwarz are more than justifiable. This dread is also revealing about the religious nature of technology; it is a mirror unto the soul; that murky metaphysical property that causes morphogenesis.

Violent technologies are simply that, machines of violent ends. But we must not begin nor end the conversation and analysis with the implement or entity, we must go further back to the mind of its creator otherwise it is merely an abstraction. As Mary Shelley has taught us in her work *Frankenstein*, Victor Frankenstein is the monster of the story, and not the creature of his making. Unlike other pieces of technology, which make some eschatological promise, death machines are created with one aim: death and destruction. What leads governments to invest billions of dollars into these technologies? What myths must the public believe in order to accept the collateral damage of thousands of women and children? In the creation of violent technologies, we see the line blur from religious to political; it becomes an unholy symbiotic relationship and addiction. The addiction to this technology is because it is high yield and low cost, at least financially speaking.

As much as I appreciate Graver's typology (Fig. 1), it is difficult to simply section off violent technologies into neat modalities. Perhaps it is even unethical to attempt such, for it allows further abstraction and thus ethical washing. As we will see in the chapter ahead, violent technologies are not created in a vacuum, but they have a way of creating a moral decision-making vacuum. Elke Schwarz is worth quoting at length here,

> This is particularly relevant in the real world, where 'the only option' is rarely just that. A failure to acknowledge the moral significance of choosing technological means might make some means that are not necessarily justified seem justified; it might make means that are not absolutely necessary seem necessary, and it might make technological tools that for whatever reason appear to be the most attractive option in a collection of available options seem like the only option. The point here is not that technological means produce entirely new moral choices, but that they carry the potential to obscure aspects of moral decision-making and thereby enable justifications for potentially greater harm.[64]

I have written this work from the perspective of a Christian theologian. There are many sins in the history of my tradition, one of which involves the acceptance and implementation of death machines. I write this book toward two ends: (1) that the public and local church would reject the making and usage of death machines. These implements are not where we should place our *faith* for security and significance; and (2) to see that prohibition (i.e., banning killer robots) is not enough. We must extend the moral circle once more to give these machines a face, or rather a faceplate. For any entity that is involved in the murky business between life and death, must have some form of moral consideration. What follows in the chapters below is an attempt to persuade the reader of the worthiness of these claims.

While violence might be a reality of nature, we must not conflate allowance with affirmation. Much like addressing the problem of evil, we must acknowledge that our epistemology is limited and often times biased. While it is easy to theorize that pain and evil are inconsistent with a loving God, there is no empirical way to prove that it is gratuitous.[65] Just like there is no way to prove all evil is gratuitous, there is, in a similar vein, no way to prove all violence is gratuitous. Here is where we need much caution and empathy. I do not idealize violence like our friend

64 Elke Schwarz, "Technology and Moral Vacuums in Just War Theorising," *Journal of International Political Theory* (2018): 5–6.

65 For a full treatment of this problem see Michael Murry, *Nature Red in Tooth and Claw: Theism and the Problem of Animal Suffering* (New York: Oxford University Press, 2011).

Foucault,[66] nor do I believe it was the ultimate will of God that violence becomes a necessary part of human history, especially the most violent act of all creation–the trial and crucifixion of Christ. While God did not create violence or the implements of violence, He created humans with free will and where there is free will, there is moral and cognitive failure. The history of violent technology begins outside of Eden; where it finishes is up to us. May God have mercy on us for the evils justified in His name.

Conclusion

Womanist theology has examined the relation between creation, violation, and God's presence. For example, Delores Williams concludes that the violation, destruction, and exploitation of land is sinful against God.[67] Karen Baker-Fletcher, commenting on this reality, notes that sin breaks the harmony of this world that God set into motion. This rebellion mutates and echoes from generation to generation.[68] I echo the warning of Williams, Baker-Fletcher, and many others, that this fall to violence is a failure of creativity and care. When we experience pain and the hopelessness of violation, we should remember there are consequences. The violation of a person's body or mind is a direct offense to the person of God. He hears every prayer and sees every crime. When humans will not cry out for justice the blood-soaked ground will. Every man, woman and child lynched from 1882 to 1964, calls out for the divine justice of God. My hope for the chapters ahead is that the reader, churchgoer, politician, philosopher, and student will see that hatred and hopelessness are the nihilistic ends when our hearts become numb to the works of our hands. The blues were songs born in pain, a creative alternative to violence, despair, and the destruction of God's creation. As you will see in the chapters below, much of the technology we are driven to create is meant to *solve* a problem. Electricity can be used for redemptive ends, the powering of a Marshall stack (guitar amplifier)

---

[66] Michael Foucault, *Discipline and Punish: The Birth of Prison*, trans. Alan Sheridan (New York: Vintage Books, 1995). Cf. Martyn J. Smith and David Hilborn, *Divine Violence and the Christus Victor Atonement Model: God's Reluctant Use of Violence for Soteriological Ends* (Eugene, Oregon: Pickwick Publications, 2016).
[67] Delores S. Williams, "Sin, Nature, and Black Women's Bodies," in *Ecofeminism and the Sacred*, ed. Carol Adams (New York: Continuum, 1993), 24–29.
[68] Baker-Fletcher, *Dancing with God*, 92.

or a pacemaker; both give life (in varying ways). However, we can also use electricity to send alternating current to an electrode attached to a person's head (i.e., the electric chair). The desires of the creators make all the difference in the ethical outcomes of the technology we design and implement.

CHAPTER 2

Of Violent Technology and Texts

> "[C]reatures eat other creatures, whether plant, animal, insects…we all participate in levels of necessary violence to feed and sustain our physical lives. A certain degree of destruction is part of the cycle of life."–Karen Baker-Fletcher[69]
>
> Violence and the sacred went hand in hand long before the rise of Akhenaten or Moses…. –John Collins[70]
>
> Suffer the children to come unto me –Mark 10:14

Introduction

Reading about violence within technology and within sacred texts is an ethical endeavor. The reader is in constant tension between overextrapolation of the historical and grammatical meaning of the text and how it translates into modern debates about the use of force in today's world. Here I would like to make a disclaimer. This chapter does not seek to dissolve the issues that surround the Bible and violence, nor does it claim to make exhaustive conclusions about the use of violence in modernity. Rather this chapter seeks to put forth an ethical reading of some of the violence within three major areas: the primeval history of

---

[69] Karen Baker-Fletcher, *Dancing with God: The Trinity from a Womanist Perspective* (St. Louis, MO: Chalice Press, 2006).

[70] John Collins, "The Zeal of Phinehas: The Bible and the Legitimation of Violence," *Society of Biblical Literature* 122, no. 1 (2003): 3–21.

Cain and Abel, the Old Testament war ethic (per genocide and rape), and the non-retaliation of Jesus in the New Testament era. This chapter will leave the reader with more questions than answers, and to a degree, that is intentional, for as I have learned from many womanist theologies, we must be careful to make conclusions about these matters (i.e., justified use of force). As a theologian, I submit all of my ideas to the judgment of the LORD who made all beings in light of His ethical design. My views and conclusions are potentially fallible and fallen to a degree. Thus, I would like to make plain my view that the violation and forceful harm of any person (not limited to moral) is a shame and transgression against the creator. That does not mean there is not a time and place for force, regrettably, but it does not make the act pristine ethically. The reader will encounter moral injury in the pages ahead, and for that I sincerely apologize, but it is a necessary work if we desire to make this world a less violent place. We must begin with our desires and justifications of both our views of violence and nonviolence, what we mean by these and how we enforce such perspectives on the gradient of violence.

Writing a theology of violent technology is a difficult task for a couple of reasons. First, the Bible does not directly address AI and robots. There are no drones in the ancient Near East, nor in the Greco-Roman world. And despite the numerous exegetical claims (or fallacies more like) that are published every year around 'end-times' in the book of Revelation, I do not believe John's apocalypse references such technology. Second, as we noted in the previous chapter, there is an entanglement between religion and technology. Religious devotion can be a wonderful and beneficial thing in the arena of technological creation. When emotions like fear and anger are processed in a healthy way, they can lead to hope, preparedness, and even awaken a desire for good. However, there is a potential impairment to these emotions; they can also lead to violence, depression, and anxiety.

A major theological question the reader should have in mind when looking at the narrative of God's story as recorded in the Hebrew Bible and New Testament, is why are violence and violation so prevalent? Could not an all-powerful, all-knowing, and all-present deity disallow or remove violence from reality? By the time you finish reading this paragraph 20 people will be sexually abused by their intimate partner.[71] In 2018, gun violence was the leading cause of death for children and teens 1–19.[72] Violence against the elderly has risen since the COVID-19 pandemic.[73] This small snapshot of violence in the present age alone posits the question, if God is all-powerful and present why does he allow such things to happen? Much like the problem of evil, this is a question we can no longer ignore. More to this, in the technological age, where there is but a membrane between violence and technology, we must reexamine this problem. To accomplish this task and frame a theology of violent technology, I will need to establish a hermeneutical framework that allows for a reading of the biblical accounts of violence with great nuance and carefulness, examining case-studies from both the Hebrew Bible and New Testament. It is the task of philosophical theology to take these readings and juxtaposed them to modern conceptions of violence and form a systematic analysis of violent technology to help give perspective to the modern reader.

The "Face of Abel" and the First Violent Implement

The study of religious violence is not for the faint of heart. Rape, murder, and infanticide are common threads that make up the tapestry of the ancient Near East. Yet there is more happening in the ancient world than brute force through the hands of primitive humans. Through studying the implements of violence throughout human history, especially in a religious context, one can see that the tool is an extension of the person's desires. The tools used to enact violence are the medium to channel the will of human characters, they go hand-in-hand. Modern reality can only manifest through a series of violent changes that are set in motion once the first rock is plunged into the back of a human skull.

> But then it was, when they were out in the field that Kayin rose up against Hevel his brother and he killed him. Yhwh said to Kayin: Where is Hevel your brother? He said: I do

[71] Centers for Disease Control and Prevention, "National Intimate Partner and Sexual Violence Survey, 2010.

[72] Centers for Disease Control and Prevention, 2020, "Leading Causes of Death Reports, 1981-2020," accessed Nov. 1, 2022, https://wisqars.cdc.gov/fatal-leading.

[73] "Violence Against the Elderly has Risen During COVID, UN Expert Warns," https://news.un.org/en/story/2021/06/1093962.

> not know. Am I watcher of my brother? Now he said: What have you done! Hark–your brother's blood cries out from the soil! (Genesis 4:2,8–10, translated by Everett Fox)

Human violence stems from enframed desires. Desire for what? What else is there to desire but sex, sapience, and immortality? This is only the beginning of the complexity, for in the earliest records of human anthropology, the will of the gods are also a part of this equation (i.e., the foundation of the enframing). Death, evil, and theodicy are universally accepted elements of ancient mythos. The ethic of the individual determines the limits to which a violent technology is operationalized.

In the story of Cain and Abel above, there is a determinism that cannot be ignored. Abel's name in Hebrew (hevel) means "puff of wind."[74] The narrator of Genesis gives us the first clue of Abel's violent fate. One that serves as a metonym for all human fate–humans are but vapor (Ecc 1:1). But there is an overarching theme in the poetics of Genesis that will thread together the rest of the Hebrew Bible–life is violence. From our violent emergence from the womb and until the pallbearers load our heavy casket upon the pneumatic lift and lower it into the earth, we are in a tension between rest and violence. The question before us is what is the force behind the violence of life and how is technology operationalized within this reality?

The story of Cain casts a long shadow on human anthropology. It is difficult to understand humanity's history with violent technology without first examining the theological and philosophical complexity of violence. Indeed, violence is multivalent, and we need to be careful about the rubric we create. In the paragraphs below, we will survey a range of modalities to help us frame violence and its usage throughout religious history. Let us start at the beginning.

---

[74] Karen Radner, *Die Macht des Namens: Altorientalische Strategien zur Selbsterhaltung* [*The Power of Name: Ancient Near Eastern Strategies for Self-preservation*] (Santag 8; Wiesbaden: Harrassowitz, 2005); Johannes Fichtner, "Die etymologische Ätiologie in den Namengebungen der geschichtlichen Bücher des Alten Testaments," [The Etymological Eitology in the Names of the Historical Books of the Old Testament] *Vetus Testamentum* 6 (1956): 372-96; Moshe Garsiel, Biblical Names: A Literary Study of Midrashic Derivations and Puns (Ramat Gan: Bar-Ilan University Press, 1991).

Cain Said to His Brother…

Why did Cain kill Abel? At first blush, one could make an inference from the text that the fratricide is a result of jealousy. Perhaps Cain is jealous that his brother's sacrifice is "regarded" and his is not. What is present within the text, and many other ancient Near Eastern parallels, is that fratricide is an outward expression of an inward devotion. Yes, it is logical to assume that Cain is struggling with inner turmoil, but that is not to say that the causation for this instance of fratricide is correlated primarily to his depression. The underlying condition that leads to the murder of Abel has more to do with religious devotion than mere aggression.

In Gen. 4:8, the phrase "Let us go to the field" is missing in the Hebrew Bible.[75] The Syriac, Latin, and Aramaic Targums include this phrase to help the reader see there was a discourse in the fields.[76] What did this conversation entail and why does it end in the murder of Abel? The conversation between Cain and Abel picks up with "the voice of blood (plural)" crying out from the soil. In the ancient versions mentioned above the reader can see an attempt to *fill the gaps* in the Cain and Abel narrative. While there most likely was not a Hebrew *Vorlage* that each version referenced, there is a common theme to explain the lacuna.[77] The attempts to smooth out the reading of the Hebrew Bible

---

[75] Harry M. Orlinsky, *Notes on the Translation of the Torah* (Philadelphia: Jewish Publication Society, 1969), 68; Radiša Antic, "Cain, Abel, Seth, and the Meaning of Human Life as Portrayed in the Books of Genesis and Ecclesiastes," *Andres University Seminary Studies* 44 (2006): 203-11; Gordon W. Wenham, *Genesis 1-15*, World Biblical Commentary, vol. 1 (Waco: TX: Word, 1987), 101-2. See also Florentino García Martínez, "Eve's Children in foe Targumim," in *Eves Children: The Biblical Stories Retold and Interpreted in Jewish and Christian Traditions* (ed. Gerard p. Luttikhuizen; kernes in Biblical Narrative 5; Leiden: Brill, 2003), 28-29; Ed Noort, "Genesis 4:1-16: From Paradise to Reality: The Myth of Brotherhood," in *Eves Children*, 104; Joel N. Lohr, "Righteous Abel, Wicked Cain: Genesis 4:1-16 in the Masoretic Text, the Septuagint, and the New Testament," *Catholic Biblical Quarterly* 71 (2009): 494.

[76] See Appendix 1.

[77] Albert Ehrman, "What Did Cain Say to Abel?" *Jewish Quarterly Review* 53 (1962): 164–67; Howard Jacobson, "Genesis IV 8," *Vetus Testamentum* 55 (2005): 564–65; M. McEntire, "Being Seen and Not Heard: The Interpretation of Genesis 4:8," in *Of Scribes and Sages*, vol. 1, ed. C. A. Evans; Studies in Scripture in Early Judaism and Christianity 9 (New York: T&T Clark International, 2004), 4–13; Michael P. Weitzman,

each give a different shade to the narrative. These attempts are not synonymous with contemporary psychoanalytic analysis in which the narrator is seeking to unearth the cognitive failure of the characters but are seeking to unpack a theological truth.

Cain is seeking to bring an offering, a "tribute" (*minha*), to win favor before the Lord. Is Cain conflating the premeditated murder of Abel with an offering to God? Perhaps this is a part of Cain's psychological profile, but there is without question the presence of anger, fear, depression, and jealousy as well. This narrative frames the history, philosophy, and psychology of violence quite well. The fields of blood that make up human history are the result of a complex synthesis of philosophical and theological worldviews. Cain's disillusioned psychological state juxtaposed with a desire to be devoted leads to a violent choice–fratricide. Yes, life is violent, but here in the episode of Cain and Abel, the first homicide, we have an example of the far end of the unacceptable spectrum.

There are elements at work in this episode of fratricide, as with others that will be examined from the ancient Near East (ANE), that must be considered in the cycle of violence and the technologies that are used as implements. Maybe it is a stretch to label a "rock" as a piece of technology. But here lies the complexities of accessing violent technology, any implement can become violent; its *telos* has now transformed. Here we are tempted, at least in modern debates about war tech, to focus heavily on the metaphysics of the implement. We might ask a question like: Can an object metaphysically change? Or What is the medium of this transformation? Design is not the only factor in the ethics of making an object, the use of such can transform it entirely. Technology is always the human's partner, and it is not neutral in that regard. These are fascinating topics and ones that I think McLuhan and Heidegger would like to see addressed, and there will be time later in this

---

The Syriac Version of the Old Testament: An Introduction (Cambridge, MA: Cambridge University Press, 1999), 74–75, 79; T. Jansma, "Investigations into the Early Syriac Fathers on Genesis," *Old Testament Studies* 12 (1958): 69–181; R. Kasher, "The Palestinian Targums to Genesis 4:8: A New Approach to an Old Controversy," in *Biblical Interpretation in Judaism and Christianity*, eds. I. Kalimi and P. J. Haas, Library of Hebrew Bible/Old Testament Studies 439 (London: T&T Clark International, 2006), 33–43; B. G. O. Kvam, "'Come let the two of us go out into the Field': The Targum Supplement to Genesis 4:8a–A Text Immanent Reading?," in *Targum and Scripture: Studies in Aramaic Translation and Interpretation*, ed. Paul V. M. Flesher (Leiden: Brill, 2002), 97–103.

work for such consideration but for now perhaps it is better to focus on why Cain murdered Abel and refused to give him a face and a voice, and the moral injury that resulted post the killing of his brother. As we can see in womanist theology and in the work of Levinas, the medium and its transformation is best understood in the faces of humans.[78]

Cain refused to give a face to his brother Abel. Yet, there is more happening here in the ethical framing of this narrative. While Cain cannot be said to be a utilitarian philosopher, it is his selfish desire, at the moment in this narrative, for God to accept his offering that leads him to turn his face from his brother. There is a process of transformation that is actualized through the medium of the rock and the murder of another human person. This does not justify Cain, nor does it remove the consequences of his sin, but it is evidence that human relationships through technological mediums are potential vessels of transformation, for restoration or for ruin. Of course, Abel is not restored to life or in any sense in this reality, but there can be no doubt that he went to a new ontological reality; a new body in the presence of God. Abel is the first *other* in the primeval history, and he will not be the last. However, the pericope of Genesis 4:8–12, the LORD gives Abel's voiceless body a voice through the saturated blood in the soil. Perhaps for the first time, Cain must face his brother and feel the weight of the death of a fellow image bearer. Levinas comments here, "[t]o be oneself is already to know the fault I have committed with regard to the Other."[79] Cain finds mercy and grace from the LORD, but he must carry his brother Abel for the rest of his life, the voice of Abel's blood will speak to him from this moment forward. We will return to this reality in later chapters in the discussion of moral injury and *post bellum* responsibility. For now, we see, in the episode of Cain and Abel, that inanimate objects can be the vehicle of both death and restoration, but that is not an argument for justifying their creation or implementation into society.

War Rape

The ethics of divine warfare cannot be ignored in a study such as this one. Nor can the reader simply download these texts into modern

[78] Emmanuel Levinas, "Simone Weil Against the Bible," in *Difficult Freedom: Essays on Judaism*, tans. Sean Hand (The Johns Hopkins University Press, 1990), 140.
[79] Levinas, *Difficult Freedom*, 20.

frameworks and expect a non-problematic ethic of violence and violent technologies will emerge. This effort requires very careful and *non-violent* or forceful readings of texts that deal with genocide: Josh 6:21; 1 Sam 15:3; Num 21:2–3; Deut 2:34, 3:6; 7:2; 13:15; 20:16, and war rape: Num 31:17–18, 35; Deut 21:10–14.[80]

There are two traditional and polarizing ways of interpreting texts like these. The first ethical perspective sees holy war and its actions as solely evil. Advocates of this view, like C. S. Cowles (i.e., *antitraditional*) and Eric A. Seibert,[81] hold that the commands for *jihad* were a cognitive failure of Moses and the Israelite leaders. For example, Cowles writes, "God does not engage in punitive, redemptive, or sacred violence. Violence and death are the intrinsic consequences of violating God's creative order; they are the work of Satan, for he was a "murderer from the beginning" (John 8:44).[82] The *traditional* reading of these texts believes that the commands *utterly destroy* are divine directives from God and are perfect and pristine ethically.[83] The typical rationale for interpretation under this ethic relates to a high view of scripture. If God says something in the Bible, or commands something, it has an inherent righteous value to it and must be obeyed at all costs to human and animal life. Eugene Merrill, and advocate for this ethical reading, writes,

> The issue cannot be whether genocide is intrinsically good or evil–it's sanction by a holy God settles that question. Rather, the issue has to do with the purpose of genocide, its initiator, and the particular circumstances of its application. We argued here that biblical genocide was part of a Yahweh-war policy enacted for a unique situation, directed against a certain people, and in line with the character of God himself, a policy whose design is beyond human comprehension but

---

[80] See William T. Cavanaugh, "Does Religion Cause Violence?," *Harvard Divinity Bulletin* 35, no. 2/3 (2007): 1–14.

[81] See C. S. Cowles, "The Case for Radical Discontinuity," in *Show Them No Mercy*, ed. Stanley N. Gundry. Zondervan Counterpoints (Grand Rapids, MI: Zondervan, 2003); Eric A. Seibert, *Disturbing Divine Behavior: Troubling Old Testament Image of God* (Minneapolis: Fortress Press, 2009); *The Violence of Scripture: Overcoming the Old Testament's Troubling Legacy* (Minneapolis: Fortress Press, 2012).

[82] C. S. Cowles, "The Case for Radical Discontinuity," in *Show Them No Mercy*, 30.

[83] M. Daniel Carroll R. and J. Blair Wilgus, eds., *Wrestling with the Violence of God: Soundings in the Old Testament*, Bulletin for Biblical Research Supplement 10 (Winona Lake: IN: Eisenbrauns, 2015).

> one that is not, for that reason, unjust or immoral. Those very limitations preclude any justification for modern genocide for any reason.[84]

If one holds this view, then it logically extends into other areas such as the treatment of slaves in Exod. 21:20–21, the dehumanization of women (Deut 25:11-12), and corporal punishment. Like genocide and war rape, these are real issues the reader will see when working through the Hebrew apparatus. In this hermeneutical approach, all holy war is a foreshadowing of eschatological ends. Longman, Merrill, and Garddraw justify this type of reading of the text through the lens of biblical scholar Meredith Kline.

The traditional reading does not get everything wrong, but neither does it get rid of all the ethical issues either. One major problem here is seeking to bind the narrative to the theology of the interpreter, especially as it relates to eschatology. That is to say, just because a biblical narrative describes a set of ethical actions, which are tied to a command (divine or not) does not imply those actions are pristine (i.e., holy war).

I am not convinced that either of the readings above are satisfactory for both a faithful reading of the Hebrew narratives and the ethical development that arise in the New Covenant through the teachings of Jesus. William J. Webb and Gordon K. Oeste in their work *Bloody, Brutal and Barbaric?* Provide a complexity thesis that I find closer to the heart of covenant theology. According to Webb and Oeste, there is a hermeneutical lens in which one can both see the ugliness of biblical holy war and see a redemptive hand at play, they label this reading the *incremental redemptive-movement ethic.*[85] Put simply, it is a theological lens for seeing the both the Old and New Covenants on a historical trajectory. It seeks to look at all sides of the biblical narratives and explore both the troubling historical contexts in which it was written (i.e., less than pristine ethics), but also "go beyond" these readings. What Webb and Oeste mean by going beyond is not a better ethic, but a rightly applied ethic. This method is not a replacement for the long cherished historical-grammatical method of reading scripture, but a subset of such.

---

[84] Eugene H. Merrill, "The Case for Moderate Discontinuity," in *Show Them No Mercy*, 93.
[85] William J. Webb and Gordon K. Oeste, *Bloody Brutal and Barbaric? Wrestling with Troubling War Texts* (Downers Grove, IL: IVP Academic, 2019), 16–17.

Reading these troubling narratives through this lens allows for a complexity thesis–there are ethical issues present in the text, but there are also redemptive elements as well. How can genocide and war rape be redemptive? I dare not posit that these disturbing actions are redemptive fully, nor divinely commanded per se in the text themselves. What is meant by a *redemptive-movement* or *incremental ethic* approach is that the best reading of troubling texts stems from reading from two horizons. On one horizon the historical context is fully accounted for. Slavery, war rape, the dehumanization of women, and the brutal treatment of children are barbaric realities, and such is not to be ignored. On the other horizon, the reader must consider the applicability (if any) of these texts for biblical ethics in contemporary culture. If the reader only considers one horizon, the text will only amount to regressive chaotic barbarianism. With both horizons in view, the reader can both understand the texts in light of other ANE war texts and practices, while also seeing echoes of these realities in our present age.

There are ugly shades to the Hebrew war texts that cannot be dismissed. One troubling text that comes to mind is Deuteronomy 21:10–14, which reads,

> When you go out to battle against your enemies, and the LORD your God hands them over to you and you take them away captive, and you see among the captives a beautiful woman, and are strongly attracted to her and would take her as a wife for yourself, then you shall bring her into your home, and she shall shave her head and trim her nails. She shall also remove the clothes of her captivity and shall remain in your house, and weep for her father and mother a full month; and after that you may have relations with her and become her husband and she shall be your wife. But it shall be, if you are not pleased with her, then you shall let her go wherever she wishes; and you certainly shall not sell her for money, you shall not treat her as merchandise, since you have humiliated her. (NASB, 2020)

This text is known colloquially within Jewish thought as the "pretty woman" text.[86] As the reader will quickly note, the ethics of the narrative do not fit within a Geneva and Hague convictional warfare ethic. Yet, the text does display both troubling ethical problems and at the same time a concern for the woman captured. Let us unpack the textual and historical evidence in order to substantiate this judgment.

Frist, notice the ethical problems latent within the text. Within four verses there is present the objectification of women, coerced sexuality, marital rape, forced marriage, patriarchy, women as property, and the overall dehumanization of the woman as a moral person. Careful exegesis is in order before moving to a systematic and theological conclusion of war rape in this pericope. In a strict sense, the passage at hand is not about the ethics of war as much as it is about domestic and civil ethics for taking prisoners of war as brides. There is an assumption in verse 10 with the penultimate sentence, "and you take them captive" that in correlation to previous clauses that God's deliverance of Israel in battle connects to his blessings of taking "pretty women." The word used here is שׁבה which means "to take captive" as a prisoner of war. This verb is used over 30 times in the Hebrew Bible in this sense, meaning it is not an isolated event within the context of Israelite war ethics and practices.[87] The focus of the passage here is on marital relations, which anticipates that there might be trouble within familial dealings in the future, and rightly so.

Another ethical issue within the text is the connection between *force* and desire (חשׁק). The desire mentioned here is overtly sexual and objectifying to the women in reference. What exactly is the text implying about these Israelite men? From a rabbinic perspective, as Ishai Rosen-Zvi has pointed out, the reader must be careful to not overload the moral

---

[86] David Stern, "The Captive Woman: Hellenization, Greco-Roman Erotic Narrative, and Rabbinic Literature," *Poetics Today* 19, no. 1 (1998): 91–127; James A. Diamond, "The Deuteronomic 'Pretty Woman' Law: Prefiguring Feminism and Freud in Nahmanides," *Jewish Social Studies* 14, no. 2 (2008): 61–85; Peral Elman, "Deuteronomy 21:10–14: The Beautiful Captive Woman," *Women in Judaism* 1 (1997): 1–13; David Resnick, "A Case Study in Jewish Moral Education: (Non)-Rape of the Beautiful Captive," *Journal of Moral Education* 33, no. 3 (2004): 307–319.

[87] In the *Qal* form: Num 21:1; 31:9; Deut 21:10; Judg 5:12; 1 Sam 30:2; 1 Kings 8:46–48; 50; 2 Kings 5:2; 6:22; 2 Chron 6:36, 38; 25:12; 28:5, 8, 11, 17; 30:9; Ps 68:18; 106:46; 137:3; Isa 14:2; Jer 41:10, 14; 43:12; 50:33.

language of "desire" reflected here. In a sense, we see an incremental (not modern in any sense) ethic trying to accommodate present understandings of sexual desire. In order to understand how that is so, some perspective on ANE war practices will better frame how Deut 21:10–14 is both regressive and incremental. This next part will not be easy to read.

The grotesque imagery required to reflect on ancient warfare helps the contemporary reader frame how the "pretty woman" text could be somewhat redemptive. Ancient warfare was not just about killing the enemy on a lateral battlefield, it was also about humiliation of an extreme nature. One common practice was to shave the pubic hair of the captives to expose their genitals.[88] This practice was a means of public shame, especially in a culture where public nudity was not normative. Another component to consider is that captives where often put into stress positions. In Mark Janzen's research on Egyptian captives of war, he notes how the iconography is a witness to the harsh positions that captives were held in. These stress positions eventually lead to torn ligaments, strained muscles, restricted blood flow, and even the dislocation of shoulders.[89] According to Janzen, the most prevalent position is the X-shape which was "one of the most humiliating and brutal of all."[90] It is not merely enough to kill and destroy men, the women and children that remain must be broken in spirit and body.

Evidenced in the passage above, slavery was also a common feature in ancient warfare. Enslavement of the captives was primarily driven by the desire for economic profit. Slavery was also a means of ensuring the suzerain-vassal treaties (SVT) was honored. An even darker shade of this practice was human trafficking which was also motivated by greed and dehumanization of the enemy.[91] The violence to the human body does not stop at war rape and slavery.

---

[88] Zainab Bahrani, *Rituals of War: The Body and Violence in Mesopotamia* (New York: Zone Books, 2008), 150.

[89] Mark D. Janzen, "The Iconography of Humiliation: The Depiction and Treatment of Bound Foreigners in New Kingdom Egypt," (PhD diss., University of Memphis, 2013).

[90] Janzen, "The Iconography of Humiliation," 61.

[91] Issac Mendelsohn, *Slavery in the Ancient Near East: A Comparative Study on Slavery in Babylonia, Assyria, Syria, and Palestine from the Middle of the Third Millennium to the End of the First Millennium* (New York: Oxford University Press, 1949), 92–94.

On top of humiliation and enslavement, it was also standard operating procedure to mutilate the body of captives. The cutting and disfigurement of hands, feet, lips, genitals, noses, and ears were tragically commonplace. Rameses III's monument at *Medinet Habu* also reveals a disturbing reality of bodily mutilation and its philosophical meaning in ANE warfare. The mortuary temple of Rameses III depicts a scene (Plate 22 and 23) where Rameses is celebrating a victory over the Libyans. These plates describe Rameses III was counting the piles of hands, penises, and scrotums that were collected from the captives; around 12,000+ hands and 12,000+ penises and scrotums. This graphic mutilation was a symbol for the utter destruction of the enemy's capacity for war and the capacity to make progenies of war.[92] This violence was not exclusive to males; pregnant women were also torn open. The Assyrian "heroic poem" VAT 13833 celebrates the victory of Tiglath-Pileser I and states, "He slits the wombs of pregnant women; he blinds the infants."[93] I think that paints a vivid image of what ANE war practices were like. So how does that compare to the Israelite version of holy war (*jihad*)?

Not all components of the Israelite war are incrementally redemptive; however, some are. Considering the brief synopsis of ANE war practices above, Israel was different, and commanded by the LORD to be such. The "pretty woman" law is just one example of an incremental ethic, that is, God implements redemptive features throughout Israel's troubled past and present actions. This does not mean, as some authors believe, that the actions are thus pristine. In fact, the Hebrew Bible records otherwise. The "pretty woman" law is one of the first historical considerations of women's rights, but that is not the end of God's requirements of Israel to reflect an *uneasy* or *reluctant* view of violence and war.

- **Stripping**: Unlike the ANE practice of stripping as humiliation, the Israelite practice was limited to outer garments (Judg 14:9; 1

[92] James Henry Breasted and Thomas George Allen, "The Historical Record of Rameses III: The Texts in *Medinet Habu*," vol. 1-II, in *Studies in Ancient Oriental Civilization*, The Oriental Institute of the University of Chicago (Chicago, IL: The University of Chicago Press, 1956), 12–14.

[93] See E. A. Wallis Rudge and L. W. King, *Annals of the Kings of Assyria* (London: British Museum, 1902), vol 1, 58; col. Iii.92–iv.4.

Sam 2:21; 23:10) and never as an act within warfare or to humiliate their enemy.

- **Tortuous Humiliation**: There is no biblical or archeological evidence that Israel ever practices similar ANE prisoner torture. Further evidenced by the treatment of slaves.
- **Slavery**: Unfortunately, chattel slavery was a practice of Israelite culture. However, it was very different from the surrounding culture. Amos (1:6–12), Joel 3:4–8, and Ezek 27:13 all condemns human trafficking. The eight commandments, "do not steal" also applies directly to humans, as we see above it was a common practice. Ignoring commands here, leads to severe punishment of Israel, a reverse holy war which was often worse than God's judgment of other nations.[94]
- **Rape and Genocide**: Again, unfortunately, these horrific acts are intertwined within the biblical narrative. In fact, they are either *permitted* or *commanded.*

Let us unpack that last point a little more. In the chapter prior to the "pretty woman" text, Deut 21 does not only grant permission to kill but *commands* it. While taking a beautiful captive was optional post-conflict, the command for total-killing was not. If we return to the traditional reading or position, there is a very clear division between the language here and thus places genocide and war rape into two different ethical categories–genocide, in this reading is pristine while war rape is not. Again, I diverge from the traditional reading and argue that they should be read together. This *incremental ethical* reading does not see the actions of the Israelites as pristine and claim that God sanctions war rape in Deut 21. Okay, but what about Numbers 31:17–18 where God does sanction total-kill and war rape? The imperatives in the verses cannot be confused for anything other than a command. Here we must make an interpretive assumption about the desires of God and the desires of humans (mostly males in this context).

William J. Webb explores this question in-depth within the context of corporal punishment and civic punishment.[95] According to Webb,

[94] See Webb and Oeste, *Bloody, Brutal, and Barbaric?*" Appendix C: Reverse Holy War (Against Israel) Was Worse. *IVP* (2019), 30–42, 5249-Appendixes.pdf.

[95] William J. Webb, *Corporal Punishment in the Bible: A Redemptive-Movement Hermeneutic for Troubling Texts* (Downers Grove, IL: InterVaristy Press, 2011), 97–118.

divine commands reflect an accommodation and, thus, an incremental ethic. If one looks at the overarching narrative of the Hebrew Bible, the picture of the LORD is of a reluctant warrior. Webb posits the following theses for this view:

- Modern readers are predisposed to seeing God as pro-war.
- God mourns violence: Ps 44:8; Lam 1:3; Ezek 2:10; Amos 5:1–2; Jer 9:10–11; Is 16:11; Jer 48:35–36.
- Killing with mercy: 2 Kings 6:8–23; cf. 2 Chronicles 28:9–15.
- David, the great military leader, is not allowed to build the Temple (i.e., bloody hands: 2 Sam 7:5–7).
- Concern for the war machine: 1 Sam 8.
- Modern readers have a lopsided reading of God's attributes (x1,000 positive and x4 negative).[96]

While Webb's theses do not explain away the violent aspects of the Hebrew Bible's depiction of God, it does paint him as subversive to war and violence generally. This theological portrait is imperative for building a theology of violent technologies. As I will argue in a later chapter, accommodation does not equal justification, as so clearly can be seen in the theories of Just War supported by many theological confessions. The major point here is that the act of war and the actions in war are often far removed from the desires of God. However, the going to war and the ethical accommodations for war are always tied to the desires of humans. This becomes more explicit when we unpack the tragic history of violent technologies and modernity's ease with their use in modern conflict. Before moving into a survey and critique of modern views of violence it is necessary to examine the images of violence and God in a case study of the New Testament.

The Non-Retaliation of Jesus

Marcion of Sinope is infamous for his dualistic belief that the God the Hebrew Bible and the God of the New Testament were distinct (i.e., demiurge). The essential idea here is that the God of the Hebrew Bible and Jesus are distinctively different in their nature and ethical systems

[96] The positive attributes: compassion, grace, slow to anger, abounding in love, faithful, and forgiving far outweigh the description of God punitive. See Webb and Oeste, *Bloody, Brutal, and Barbaric*, 287–313.

(i.e., the LORD is wrathful and jealous while Jesus is meek, mild, and peaceful).

Yes, there is a presence of violence in the apocalyptic literature of Revelation which depicts one final holy war with Jesus as the primary warrior. The gentle and humble person of Jesus seems to disappear from the pages of John's vision in Revelation. Again, as with texts of war-rape and genocide, the reader must be ethical in the reading of such violent text. Properly understanding the metaphor of the violent imagery of Jesus in Revelation puts the reader in tensions with understanding the person of Jesus as a minister. I submit to the reader that Jesus promotes *antiviolence* or alternative responses to worldly violence that either require passivity or protest. How did Jesus challenge his followers in this eclectic approach?

The ministry of Jesus is profoundly focused on the *other* and those who were considered enemies of God. In the book of Luke there is an episode where Jesus opens the scroll of Isaiah and interprets the prophet's message in light of his incarnation.

And He unrolled the scroll and found the place where it was written:

> The Spirit of the Lord GOD is upon me, Because the LORD has anointed me to bring good news to the afflicted; He has sent me to bind up the brokenhearted, to proclaim liberty to captives and freedom to prisoners; To proclaim the favorable year of the LORD And the day of vengeance of our God; To comfort all who mourn,
> And He rolled up the scroll, gave it back to the attendant, and sat down; and the eyes of all the people in the synagogue were intently directed at Him. Now he began to say to them, "Today this Scripture has been fulfilled in your 1hearing." Luke 4:18–21 (NASB, 2020)

Here is the hermeneutical *rub*, so to speak, Jesus reinterprets the *Torah* and gives a new vision for the people of God. His message to proclaim peace and love to the outsiders and the welfare of all nations ironically rouses a collective violence against him that leads, eventually, to death on a cross in Palestine. The incarnation of Jesus perplexes worldly systems of power and the means and implements that are required to enforce such. We get the fullest sense of an anti-violent God through the lens of Jesus' public ministry. We must frame this properly before we can critique violent technologies faithfully.

There are two thoughts reflected in Jesus' categorical moral commands about nonretaliation and the love of enemies. One set is in the book of Matthew 5:27–49 (pronouncements on the mountain) and Luke 6:27–36 (the plain). Both sets are situated in a geographical and political context familiar to a first-century audience. My work here is to recapitulate those teachings in light of the current gradient of violence and violent technologies. First, let us ask: what is Jesus' perspective on violence?

The two sets of teaching mentioned above vary slightly in what and how they empathize with the person of Jesus. Luke's account (lower setting) presents Jesus' teaching on a plain. From a close social distance, Jesus engages the social inequalities and injustices present in the material world at that time. Matthew's account (higher setting) varies because Jesus is pictured teaching on a mountaintop. This recalls the imagery of Moses dictating the Ten Words to the people of Israel. In addition to this variation, Matthew also records a more general and third-person perspective (i.e., "Blessed are the poor in spirit"). Luke, on the other hand, records in his gospel a direct connection between Jesus and the crowd, "you will be full."

Luke's account chooses to emphasize that those without power (i.e., widows, children, women, and the lame) do not have to resort to the violent means of those around them. Jesus' teaching gives two imperatives in the Lucan account: (1) love your enemies and (2) bless those who persecute you (vv. 27–28). Okay, but who is the enemy in question here? The root word for enemy here is ἐχθρός ("enemy or hostile"). It could refer to the Romans, whose systems allowed for violent retribution.[97] The enemy could also be the gentiles or a general reference to the complexities of living within a village community. While this is not a full-scale rejection of the system of Hellenistic and Roman reciprocity (there are positive and negative elements), the categorical imperative in the Lucan account does refuse negative reprisal. Why?

[97] See Craig Anderson. *Roman Law for Scots Law Students* (Edinburgh: Edinburgh University Press, 2020), 158–279. Cf. Alan Kirk, "'Love Your Enemies,' The Golden Rule, and Ancient Reciprocity (Luke 6:27–35)," *Journal of Biblical Literature* 122, no. 4 (2003): 667–86; Richard Horsley, *Jesus and the Spiral of Violence Popular Jewish Resistance in Roman Palestine* (San Francisco: Harper & Row, 1987).

Simply put, violence leads to violence. Measure for measure is an escalation of force that does not end with peaceful resolve, nor does it achieve the desired end of bringing peace to the victim. What about the account in Matthew?

The Gospel of Matthew gives a slightly different emphasis on the teachings of Jesus about non-retaliation and the love of enemies. Matthew's account also gives Jesus' declarations that express an antithetical ethic of non-violent retaliation. The first of three couplets refers to *lex talionis* ("eye for an eye and a tooth for a tooth"). In a sense, Jesus is turning this teaching on its head; however, that does not mean Jesus is saying to permit all evil and violent actions without some measure of interference or retaliation. The emphasis in this context is to go beyond the Mosaic law; while measured retaliation was permissible, it will not break the chain of violence. In verses 38–42, Jesus substitutes a principle of revenge with one of self-restraint. There are a few examples given in the teaching: resisting the evil one, not striking back for injury or insult, surrendering one's outer cloak, and going the extra *mile*. Why respond in this manner? Violence for violence does not ultimately bring restoration. Nor does it fully bring justice, which is at the core of God's person and the primary reason why Jesus becomes the *face* of God–to bring *shalom* to the relationship between creation and creator.

What is fascinating here is that Jesus' categorical imperative is aimed directly at the desire to justify violent resistance. There was a great deal of tension between the ideologies of the Roman cult and Jewish orthodoxy to say the least.[98] The history of interpretation also posits that these passages above can be interpreted in a multitude of ways, leading to multiple theories of just war (JWT) or pacifism. Indeed, but as noted above, Jesus' view of non-retaliation does not automatically lead to a view of nonviolence.

---

[98] For similar logic in the early church fathers see Augustine, "Epistle to Marcellinus," no. 138 (ad 412), in Philip Schaff (ed.), *Nicene and Post- Nicene Fathers* (Buffalo: The Christian Literature Company, 1886), vol. 1, 2.13; Thomas Aquinas, *Summa theologica* (London: Blackfriars, 1963), bk. I-II, 40. Cf. Martin Luther, "Temporal Authority: To What Extent it Should be Obeyed," in Walther I. Brandt (ed.) *Luther's Works: Christian In Society II*, vol. 45 (Philadelphia, PA: Fortress Press, 1962), 88; John Calvin, *Institutes of the Christian Religion,* trans. Henry Beveridge (Grand Rapids, MI: Eerdmans, 1989), 4.20.20; Reinhold Niebuhr, "Why the Christian Church is Not Pacifist," in *Christianity and Power Politics* (Hamden, CT: Archon Books, 1969), 1–32.

There are a few episodes in the NT that are troubling for those who hold that Jesus promotes a view of complete nonviolence. The first episode is found in the clearing or cleansing of the temple in Luke 19:45–46. While Luke's account leaves out the violent actions of whipping, table turning, and scattering coins (cf. Mark 11:15–17; John 2:13–22), there is still a witness to an aggressive outpouring of anger. What provokes God to anger is that the religious leaders are misusing the temple. So, violation of sacred space leads to violent force in this episode.

In another related NT text, Luke 22:35–38, Jesus tells the disciples to buy a sword if they do not already have one. "But now, let anyone having a purse take it and likewise a pouch. Anyone not having a sword should sell his coat and buy one." Why do the disciples need swords and to what degree are they allowed to use them in response to force? Context is everything. Here the disciples misunderstood Jesus and the episode of the two swords is no different. The irony of the context must not be overlooked. The opposition to Jesus' ministry is large scale, including the Roman government. Do the disciples think two swords are enough to engage in resistance against the forces of the government? The disciples are to leave their outer garment (i.e., sustenance) for a piece of metal. Times are desperate. Jesus in verse 37 quotes a portion of Isaiah 53:12 to help the disciples see that his ministry is coming to a violent end and the prophecy will be fulfilled. However, the disciples miss his intended meaning. In response to Jesus, the disciples said, "Lord, look, here are two swords." And He said to them, "It is enough" (v.38). The passage closes with a rebuke (cf. cf. Deut 3:26; also, Gen. 45:28; Ex 9:28; 1 Kings. 19:4; 1 Chr 21:15). The desires of the disciples are for Jesus to conquer violence with violence. But the teachings and way of Jesus, per the cross, show that measure for measure never works. As we see in the message of Jesus, in particular the Sermon on the Mount, there is another way, or what Walter Wink labels the "third way."[99] Wink's short work highlights Jesus' vision to meet violence with restoration. Resorting to violence was more about a lack of creative response, and as we have already mentioned it further continues the cycle. Breaking the cycle means seeking the irredeemableness of violent retroaction. As Wink notes, the teaching of Jesus is to show that all humans bear the image of God and that true

[99] Walter Wink, *Jesus and Nonviolence: A Third Way* (Minneapolis: Fortress Press, 2003).

justice and not be accomplished through means that only further creation new oppression and injustice.

The last example, and potentially the most disturbing, is Jesus as a warrior in the book of Revelation. How do we reconcile the peaceful and nonviolent religious movement of Jesus with the apocalyptic warrior Jesus in the book of Revelation? In the *traditional view*, the imagery is justified because it reflects the same pristine ethic of war as found in the OT. War imagery in passages like Matt 24:3-32, Mark 13:3-27; and Luke 21:5–28 present a similar OT warrior portrait of the person of Jesus. Scholars like Tremper Longman state, "Jesus in the New Testament is no less violent than the revelation of God in the Old Testament."[100] This logic is not convincing. Ethically problematic statements in the present do no wash ethically problematic actions in the past. In Rev 19, John recounts in his vision that Jesus rides into the scene of the final battle on a white horse welding a sword. The imagery used throughout the book of Revelation reflects war language. The spoils of war go to the victor (Rev 2:7, 11, 26; 3:5), the followers of Jesus are "conquerors" (νικάω) and encouraged to follow the sword-wielding Jesus (Rev 2:12). What is even more troubling is that Jesus is depicted as fighting against congregations in Asia Minor prior to the holy war in Rev 19. Thus, based on the ambiguity in the categorical imperatives of Jesus' teaching on nonretaliation, the accommodation for war in the OT, and the warrior imager in Rev, it seems logical, in the *traditional view*, that war, violence, and holy war are pristine ethically. No!

While I follow Webb and Oeste in reading the imagery of the warrior Jesus as metaphorical, doing such does not resolve all the ethical issues at hand. The non-literal reading portrays a consistent vision of Jesus overcoming the enemy in ways that are not physically violent. In Rev 19, there are no chariots, military equipment, or massive armies of soldiers, just Jesus and a sword (or a metaphor). The sword is in the mouth of Jesus, for he is the "word of God" (Rev 19:13). The imagery here is that through the words of Jesus the battle is won.[101] As we will see in the chapter on just war theory, the traditional view has justified the Christian

[100] Tremper Longman, "A Response to C.S. Cowles," *Show Them No Mercy*, 58–60.

[101] David L. Barr, "Doing Violence: Moral Issues in Reading John's Apocalypse," in *Reading the Book of Revelation: A Resource for Students*, ed. David L. Barr, Society of Biblical Literature Resources for Biblical Study 44 (Atlanta: Society of Biblical Literature, 2003), 97–108.

shedding of blood since the time of Augustine. Because there is the presence of violence (at times even endorsement) within the Scriptures and no matter-of-fact denunciation of warfare, it is far too easy for interpreters to wield the Bible as a vehicle for political power. Why is the default reading justification of violence and not the renunciation of violence? This goes back to our initial assessment of violence and its implements in technology; it is always steeped in the desire to force one's worldview. Well, how do we reconcile the images of violence above and their misapplication through the ages?

Various interpreters stop reflecting on the violence of the Bible when they get to the episode of the cross. Others skip ahead to the violent imagery of Revelation. Both approaches are problematic. Either violence in the Bible is seen as supporting or condoning such, or it is read through a metaphorical lens. To stop at the narrative of the cross is to miss the peace of the resurrection and the restoration of the image of God, which ironically is the point of the metaphors in Revelation.[102] In the book of Revelation, there are three sins that call out for justice: economic tyranny, sorcery or idolatry, and murder.[103] Much like the teaching ministry of Jesus, the stories John used in his revelation illustrate and point to a deeper theological meaning.[104] Unfortunately, how these metaphors translate and shape the imagination of the reader can lead them astray and into unethical renderings of the text (i.e., violence). Metaphors are powerful mechanisms in the minds of humans, in fact, according to George Lakoff and Mark Johnson, they can transform how we understand everything else.[105] The violent metaphors in Revelation are no exception, "swords," "conquering," and "battlefield" language are

[102] Cf. G. K. Beale, *A New Testament Biblical Theology: The Unfolding of the Old Testament in the New* (Grand Rapids, MI: Backer Academic, 2011), 462–65.

[103] Grant Osborne, *Revelation*, Baker Exegetical Commentary, (Grand Rapid, MI: Baker Academic, 2002), 658–59.

[104] See David L. Barr, "Doing Violence: Moral Issues in Reading John's Apocalypse," in *Reading the Book of Revelation*, 97–108; "Towards an Ethical Reading of The Apocalypse: Reflections on John's Use of Power, Violence, and Misogyny," in Society of Biblical Literature 1997 Seminar Papers 36 (Atlanta: Scholars Press, 1997) 358–73; Brian K. Blount, *Revelation: A Commentary* (Louisville: Westminster John Knox, 2009) 1-5; William Klassen, "Vengeance in the Apocalypse of John," *Catholic Biblical Quarterly* 28 (1966): 300–311.

[105] George Lakoff and Mark Johnson, *Metaphors We Live By* (Chicago: University of Chicago Press, 1980), 5.

prevalent. Also, the metaphors used conflict with one another at times, for example, the lion and the lamb language in chapter 5. I agree with Susan E. Hylen, that the employment of these metaphors helps the reader deal and wrestle with the attributes of God, the demand for justice, and restoration of dignity to God's image bearers.[106] The book of Revelation, under this reading, is entirely an ethical project. It puts theological and political tension before the reader, forcing them to question their systems of logic, application of theology, individual values, and ultimately their participation in conventions of violence. The message of Revelation is about the reality that Christ conquers evil, not with a battle, a physical sword, or other conventional mechanisms of human violence. John uses metaphors of violence to expose the reader's imagination, which is where violence both begins and ends.

Now we can turn towards the *face* of these systems and ask, like Jesus, whether these systems will lead to violence or peace? In many surveys of violence, the literature under analysis usually starts in the late modern age (c. 1700s–1900s). It is common in the literature review to see the discussion about violence begin with thinkers like John Locke, Karl Marx, Fredrick Engels, Maurice Merleau-Ponty, Georges Sorel, Simone de Beauvoir, and so on.[107] Yet, as we have already seen above and in the last chapter, the political and religious nature of violence is by no means a modern creation. The major difference between the late modern age to the present is that the religious framework is absolved into the political, and without a proper acknowledgment of this reality I would add. Therefore, the remaining components of this chapter will not revisit what has already been firmly analyzed in the secondary literature of the thinkers above. What is needed now is to further investigate a theological framework to address violent technologies that builds off the logic found in chapter 1.

The technologies that humans create are not always innately violent, but some are, unfortunately. As we will see in the next chapter, the international investment in violent systems has a long paper trail. Before

---

[106] Susan E. Hylen, "Metaphor Matters: Violence and Ethics in Revelation," *Catholic Biblical Quarterly* 73, no.4 (2011): 777–96.

[107] See Charles Finlay, "Hannah Arendt's Critique of Violence," *Thesis Eleven* 97 (2009): 26–45; Slavoj Žižek, *Violence: Six Sideway Reflections* (New York: Picador, 2008); and Elizabeth Frazer and Kimberly Hutchings, *Violence and Political Theory* (Medford, MA: Polity Press, 2020).

we get into the *what* of these systems, it is also important to understand the *why*. This question was posed by Hannah Arendt in her seminal work *On Violence* in 1970, but also further investigated by the brilliant work of Elke Schwarz in her book *Death Machines: The Ethics of Violent Technologies*.[108] Schwarz's work, in a compelling and convincing presentation, gives witness to why modernity makes peace with violent weapon systems, even justifying their existence. She labels this phenomenon as "biopolitical conditioning" Schwarz writes,

> [B]iopolitics refers to the institutionalized mechanism and discourse of power over the body and biological functions at the individual and population level, whereby political government and life government are folded in with one another for the administration of politics. In the master and meta-mandate to secure the health, prosperity, survival, and progress of population, biopolitics is inseparably entwined with concern and practices of control, prediction, and prevention.[109]

Again, this acceptance of biopolitical control is a very religious concept, one that fits within the theological framework of idolatry. In the ancient world, idolatry was focused on two concerns, security, and significance. Society today may not erect Asherah poles on top of hills, but we still struggle (dare I say, rebel) to control and determine what is seemingly chaotic. Autonomous weapon systems (AWS) do not appear *ex nihilo,* nor do they emerge into society *tabula rasa*. The question that Jesus would have us ask is "why have we made peace with systems of death and destruction?"

The current biopolitical conditioning, to borrow a phrase from Schwarz, is complex and deeply religious. David F. Noble, in his work *The Religion of Technology*, examines thecomplex relationship between religion and technology. Modern technology, according to Noble, is born through the ideology of monasticism.[110] The eschatological views of the

---

108 Elke Schwarz, *Death Machines: The Ethics of Violent Technologies* (Altrincham Street, Manchester: Manchester University Press, 2018).

109 Schwarz, *Death Machines*, 6.

110 David F. Noble, *The Religion of Technology: The Divinity of Man and the Spirit of Invention.*

writers of the Middle Ages, those like Joachim of Fiore,[111] recapitulated a vision of a new Eden that would usher in for God's elect through the mechanisms of technology. It was their devotion that led them to engineer the pathway to modern technological society. What began as a means to worship, study, and gardening quickly transformed into corruption by distorted desires? As George Ovitt Jr. recalls, "The poor monks who once maintained themselves through manual labor became feudal lords who supervised the work of others."[112] The notion that medieval thinkers were working to bring about the new millennium juxtaposed to a view that the source of sin is the rebellion of nature provides a clear picture of why this *evolution* of technology ultimately ends with the creation and justification of violent technologies. From the visions of Joachim of Fiore, continental reformers Johann Andreae and John Comenius, the conquering of nature by Francis Bacon (cf. *Novum Organum*), to the buzz cuts at NASA, the quest for a new Eden and new Adam have fueled an endless religious thirst for power and ultimately apocalypse.

One's view of the millennium or eschatology informs the ontology and teleology of the technology created. Technology always becomes apocalypse. Nowhere is this more prevalent than in the analysis of the justification of violent technologies, particularly atomic weapons in recent memory. Leo Szilard, physicist who patented the nuclear fission reactor, believed that atomic energy had a salvific nature to it. He writes, "if I wanted to contribute something to save mankind, then I would probably go into nuclear physics, because only through the liberation of atomic energy could we obtain the means which would enable man not only to leave the earth but to leave the solar system."[113] Is this not reflective of the eschatology of billionaire Elon Musk in his SpaceX program?

Robert Oppenheimer, head of the Los Alamos Laboratory and one of the creators of the atomic bomb, named the first atomic test Trinity.

[111] See Dojčin Živadinovič, "The Origins and the Antecedents of Joachim of Fiore's (1135–1202): His Historical-Continuous Method of Prophetic Interpretation," *Andrew University Seminary Studies* 56, no. 2 (2018): 330–31.

[112] George Ovitt Jr., *The Restoration of Perfection: Labor and Technology in Medieval Culture* (New Bunswick, NJ: Rutgers University Press, 1986), 153.

[113] Quoted in Richard Rhodes, *The Making of the Atomic Bomb* (New York: Simon & Schuster, 2012), 25.

It is unclear as to why Oppenheimer would give such a blasphemous Christian code name to such a test. However, what is clear is that for Oppenheimer the atomic bomb clearly symbolized the realization of religious devotion.[114] The successful test and later deployment of atomic bombs triggered a pendulum swing in the apocalyptic vision of religious thinkers, particularly in fundamentalist perspectives. The vision of the new Eden in this millennium was destroyed along with the horrific explosion at Hiroshima. After these events, atomic energy was now an implement of Armageddon.

In as little as forty years, the concern for the use of force (i.e., nuclear technology) now occupied not only religious thought but also foreign policy. The Cold War was a witness to the religious fervor to secure power through the most violent technologies available at the time. During these times over sixty-five thousand warheads were amassed.[115] As noted in the last chapter, religion and technology are entwined. The religious experience of the technologist is the enabling mechanism that moves the interlocuter from the creation to the deployment of implements. Roger Bacon, Isaac Newton, Robert Boyle, Athanasius Kircher, and many others confirm that the history of technology is far from a neutral field. And while the 19th century sought to segregate religion from technology, what actually happened was the exchanging of one faith in favor of another. Here lies the crux of our modern predicament with violent technologies, is there room for violent technology within the Christian worldview?

How does this journey speak to the religious technology of violent implements? It is critical that the reader understand that the arguments or rather justification for death machines has more to do with rhetoric than philosophy. For example, in February 2022, Senator Lindsey Graham gave a forceful speech during a Senate Judiciary Committee hearing about the consideration of scaling back the US drone program. Notice, in particular the religious language in an exchange between Sen. Graham and Hina Shamsi, National Security Project at the American Civil Liberties Union (ACLU),

---

114 Oppenheimer in an interview with General Leslie Groves recalls, "Why I chose that name is not clear, but I know what thoughts were in mind." Quoted in *Atomic Bomb*, 572.

115 "Table of Global Nuclear Weapons Stockpiles, 1945-2002," Natural Resources Defense Council. Available from http://www.nrdc.org/nuclear/nudb/datab19.asp. Accessed September 13, 2011.

> Mr. Graham: "Miss Shamsi, do you believe that if Al-Qaeda or ISIS could attack the American people they would?"
> Miss Shamshi: "Senator Graham, what I think is important..."
> Mr. Graham: "No, what is important is that you answer the question first then give an explanation. It is a simple question.
> Miss Shamshi: "The reason I am having a hard time answering that with a yes or no is because..."
> Mr. Graham: "That is all I need to know, because no one in this world should have a hard time answering that with a yes or no. If you can't answer that yes or no, you have no idea what you are talking about. You are living in a dream world. Because I will tell you right now ma'am, if they could kill us all they would. The only reason three thousand Americans were killed on 9/11 is because they could not find a way to kill three million of us."[116]

The rhetoric within Senator Graham's discourse frames the American perspective and justification of the political and religious acceptance of death machines. This is further substantiated by the long financial paper trail that Washington and the Pentagon have created to support their biopolitical and religious frameworks. Violence through technology is not only acceptable, but necessary. Even evangelical perspectives on violent weapon systems accept this ethical conditioning. The Ethics Religious Liberty Commission's "Artificial Intelligence: An Evangelical Statement of Principles" states the following about AI and war,

> We affirm that the use of AI in warfare should be governed by love of neighbor and the principles of just war. The use of AI may mitigate the loss of human life, provide greater protection of non-combatants, and inform better policymaking. Any lethal action conducted or substantially enabled by AI must employ human oversight or review. All defense-related AI applications, such as underlying data and

[116] "'You're Living in A Dream World': Graham Claims Witness Downplaying the Threat of Terrorism," The Hill YouTube Channel, *YouTube*, video, February 2022, https://www.youtube.com/watch?v=y8m-8EUhn4I, (accessed Nov 17, 2022).

> decision-making processes, must be subject to continual review by legitimate authorities. When these systems are deployed, human agents bear full moral responsibility for any actions taken by the system.
> We deny that human agency or moral culpability in war can be delegated to AI. No nation or group has the right to use AI to carry out genocide, terrorism, torture, or other war crimes.[117]

The justification for AI in warfare, even when the principles of just war theory have been applied, must be challenged. Our brief exercise into the thinking and application of religious violence demonstrates the membrane between desire and creating implements of death. How can we love our neighbor and destroy their land, loved ones, and livelihood? How can Christians claim to hold the categorical imperative of non-retaliation while also accepting the logistical, biomedical, psychological, and physical harms of implements of violence and war?

The harsh reality of this theological investigation is that it does not require the rigors of biblical hermeneutics to see that society should reject the creation and implementation of violent technologies. As evidenced in the statement of principles above, this rejection is not normative. Why? Because as Jennifer Terry has astutely documented in her research, the West is *attached to war*.[118] While this attachment became apparent in the wake of the events of 9/11, a theology that allowed for such an attachment was present since the just war theory of Augustine. However, the Bush administration took full advantage of this biopolitical knowledge and leveraged it toward the campaigns in Iraq and Afghanistan. President Bush claimed, "Americans are free people who know that freedom is the right of every person and the future of every nation. The liberty we prize is not America's gift to the world; it is God's to humanity."[119] The theological framing here accommodates for

[117] "Artificial Intelligence: An Evangelical Statement of Principles," Article 10: War, *ERLC*, April 11, 2019, https://erlc.com/resource-library/statements/artificial-intelligence-an-evangelical-statement-of-principles/, (accessed Nov 17, 2022).

[118] Jennifer Terry, *Attachments to War: Biomedical Logics and Violence in Twenty-first-century America* (Durham, NC: Duke University Press, 2017).

[119] George W. Bush, "State of the Union Address," *Washington Post*, January 28, 2003, https://www.washingtonpost.com/wp-srv/onpolitics/transcripts/bushtext_012803.html (accessed Feb 23, 2023).

calculated loss of life, even making it pristine when outsiders threaten the life, liberty, and resources of those in positions of power.

A theology of violent technologies does not simply allow for death machines but makes them a necessary evil considering the potential benefits and ultimately *hope*. The promise of war-building that it will bring safety, security, and developments in regenerative medicine, is a deception. The promise of AI and robots in war-time applications is that it will make war "cleaner" and provide humanitarian aid, but is this a promise that mathematical models can make? The answer is no. Mathematical models can be the basis of ethical malpractice (cf. *bugsplat*)[120] and even propagate the very harm they claim to prevent.[121] In fact, according to the research of Ned Debos, because of the *fear of reprisal* that goes along with the creation of these technologies, it leads to more conflict not less.[122] Therefore, even in purely defensive systems, as in one that I have first-hand experience with like the Phalanx system,[123] there is still a great ethical cost to violent technologies that violates the personhood of combatants and civilians alike. As we will see in the next chapter, the West sacrificed the privacy of self and neighbor, the lives of thousands of women and children, destroyed present and future economies in the name of political salvation, biomedical enhancement, and the ethically washed American dream.

It is not that we should not have hope or attachment, in fact, these are deeply religious concerns and mechanisms for not only survival but health. Attachments to the implements of war are close to the theology of hope put forth by theologian Jürgen Moltmann in *Theology of Hope*,[124] which is primarily oriented toward the future (i.e., eschatological). As

---

[120] John Emery, "Probabilities Towards Death: Bugsplat, Algorithmic Assassinations, and Ethical Due Care," *Critical Military Studies*, https://doi.org/10.1080/23337486.2020.1809251, (accessed Feb 23, 2023).

[121] Cathy O'Neil, *Weapons of Math Destruction: How Big Data Increases Inequality and Threatens Democracy* (New York: Crown Publishing, 2016).

[122] Ned Debos, *Ethics, Security, & The War-Machine: The True Cost of the Military* (New York: Oxford University Press, 2020).

[123] "Phalanx Weapon System," *Raytheon Missiles & Defense*, www.raytheonmissilesanddefense.com/what-we-do/naval-warfare/ship-self-defense-weapons/phalanx-close-in-weapon-system (accessed Feb 23, 2023).

[124] Jügen Moltmann, *Theology of hope: On the ground and the implications of a Christian eschatology*, trans. J. D. Roberts (Minneapolis: Fortress Press 1967).

sufficient critics of this perspective of hope have pointed out,[125] this framework is privileged at best, if the individual has time, power, and position. Hope is not unilateral, it moves persons through their past, present, and future desires. Hope also requires attachments to objects, persons, or things. In the survey of the politics and ideologies imbedded in death machines, the reader will see that the *hope* in the West is very much tied to the military-industrial complex. It is the result of an insecure attachment at the limbic level.

We have only touched the surface of the relationship and theology of these technologies. The waters ahead are deep and murky. Theologies that accommodate war machines, implements of death, and the like, making them pristine in the face of scriptural ambiguity, do not belong in Christian doctrine. More to this, while Jesus does not explicitly call for nonviolence or hard pacifism, he does speak about attachments that lead to eternal death. Jesus states the following in the Beatitudes: "And if your right hand is causing you to sin, cut it off and throw it away from you; for it is better for you to lose one of the parts of your body, than for your whole body to go into hell" (Matt 5:20, NASB, 2020). The emphasis in this discourse is not on the implements or the object of desire, but the human as a vehicle of desire for either ruin or restoration. Addressing the religious, political sect of the Pharisees, Jesus gives a new plane for understanding righteousness. It is not merely about the letter of the law, but the spirit of the person. It is this theological lens that makes it difficult to see Jesus of Nazareth co-signing usage of drones, or the ERLCs statement of principles article on AI and war. Technologies are value laden, and those values are engineered and coded by human spirits and desires. This gets to the question posed by Judith Butler in *The Force of Non-violence,* "Can violence remain a mere instrument or means for taking down violence–its structures, its regime–without becoming an end in itself?"[126] Perhaps it is logical to agree that Jesus, like Butler and Walter Benjamin, would question the instrumentalist framework, because it too simply abstracts the persons involved. If violence is a wine, the West is in a continuous state of intoxication. We cannot overcome this addiction by consuming and producing more of the substance that enslaves us. So, where must we begin?

---

[125] Katie Ernst, "Terror of Hope: The Integration of Attachment Theory and Womanist Theology," *Pastoral Psychology* 66 (2017): 201–212.

[126] Judith Butler, *The Force of Non-Violence* (New York: Verso Books, 2020), 13.

The pathway forward, one of healing, is forged through first realizing that there are many lies we believe in the West to justify violent technology and the massive wake of destruction it leaves behind. In theological terms, this is known as μετάνοια or repentance (lit. "to turn or a change of mind"), a turning from one way of life toward another. Indeed, but where must we turn? I suggest to the reader that we must *turn* from an individual concern for safety and security toward the face of the *other*. The making of violent technologies is an egregious act of individualism. Its justification is rooted in the dehumanization of the enemy. If that was not bad enough, developing and employing such machines of death requires strata and strata of biopolitical and often religious indoctrination. We must turn to *face* the machines of death we have created. In the recent literature on animal and robot rights, this is labeled the "relational turn," that is, to consider entities that may not share our face, DNA, hierarchy on the ecological chain of being. However, there are relational values at stake, and a change of mind is needed.[127]

As mentioned above, the Christian arguments for pacifism must go far beyond speculations if they may wish to base their rationale on the heuristics of the Sermon on the Mount. The Bible, and Jesus' teachings do not strictly command nonviolence. Perhaps one reason for this is because of the precariousness of life and the humanistic temptation to deploy legalistic hermeneutics. Jesus does not specify *who* or *what* is justified to kill, most likely because he desires us to see the alterity of the other and wrestle with the precariousness of life. Humans and animals exist in interdependence, both to one another and to the environment. What human history has shown us is that we seek out ontological distinctives to determine how to employ our social and political frames. Again, Judith Butler is helpful here in her political philosophy of the framing of war (cf. enframing). Nonviolence is an issue of equality, and it is a forceful commitment to seeing the *face* of those whom we are tempted to damage or destroy.[128] The societal and political frames that

---

[127] See Mark Coeckelberg, *Growing Moral Relations: Critique of Moral Status Ascription* (New York: Palgrave Macmillan, 2012). For a critical analysis of this perspective toward robots see David Gunkel, "The Relational Turn: Thinking Robots Otherwise," in *Social Robotics and the Good Life: The Normative Side of Forming Emotional Bonds with Robots*, eds. Janina Loh and Wulf Loh (Bielefeld, Germany: Transcript Publishing, 2022).

[128] Judith Butler, *Frames of War: When Is Life Greievalbe?* (New York: Verso Books, 2016).

are forced on us through our contextual frame will determine which life is grievable to lose and which life is not grievable. Christians in the West might say killing a human fetus is grievable but killing a Muslim in the post 9/11 war campaigns was not grievable. This approach is not an egalitarian vision or commitment to the value of life.[129] As mentioned in the previous chapter, violence takes the shape of many modalities, but violent technologies become a vacuum in which the gradient is dissolved. The question of violence and nonviolence is not about peace and war, but about the path to navigate the political frames through which we see the precariousness of life. It is about resisting the act of establishing faux ontological boundaries that either bind or unbind the collective consciousness.

Justice and Violent Tech

One of the many reasons I value a womanist perspective in reading about violence and technology is because the treatment of such subjects is never left in the abstract. Violent tech may be a nebulous term, but it leaves an embodied trail of pain and suffering. It is a more holistic treatment in the systematic treatment of theological texts and ethical outcomes. Violence and technology saturate our culture, and of utmost concern for this study are the weak and powerless: children. I appreciate the words of Albert Camus in his discourse on the relation between the believer and unbeliever concerning violence, "When a Spanish bishop blesses political executions, he ceases to be a bishop or a Christian or even a man; he is a dog just like the one who, backed by ideology, orders that execution without doing the dirty work himself. We are still waiting, and I am waiting, for a grouping of all those who refuse to be dogs and are resolved to pay the price so that man can be something more than a dog."[130]

In my own theological tradition (Baptist) so much emphasis is placed on the suffering and death of Jesus in the crucifixion narrative. The

[129] Cf. Butler, *The Force of Nonviolence*, 61.

[130] Albert Camus, "The Unbeliever and Christians," in *Resistance, Rebellion, and Death*, trans. Justin O'Brien (New York: Random House, 1960), 54.

suffering on the cross is a profound piece of Christian theology.[131] It is part of the answer to the call for divine justice and for appeasement (Gen 3:7; Isa 1:4; Jer 2:22; Rom 3:23). In the views of atonement theory, the objective of Christ's death varies. The classical view (i.e., Ransom) emphasizes the cross as the moment where Jesus defeats sin, death, and ultimately the Satan.[132] Many medieval theologians took Anselm's juridical theory, the cross is about God's honor, and tied it to the classical view (i.e., Hugh of St. Victor and Franciscan Bonaventura). Peter Abelard (d. 1142) was the first to promote a different perspective of the cross, one of exemplarism. In this view, the cross is not about atonement per se, but about behavioral modification. Abelard wrote, "Christ died for us in order to show how great was his love to mankind and to prove that love is the essence of Christianity."[133] In contradistinction to these moral influence viewpoints of atonement, Karl Barth put forth the Universal Reconciliation theory. Essentially, this view states that in the suffering and death of Jesus, all humankind will be reconciled through Christ as humankind's representative.[134]

The last theory, one that I was taught in my tradition, is the Penal Substitution view. The cross, in this view, is not about God's honor but about law. Christ's death on the cross is the perfect substitute for the death and destruction of mankind since the fall of Adam and Eve. Readers of Augustine, Calvin, Luther, John Owen, and many others will see the connection in their theology to this view of atonement. It is not only a crucial (hence *crux*) piece of reformed theology but also in many practical ways, the sole teaching of many of my fellow pastors and congregants. To be fair, the Gospels are hyper focused on the cross and the final week of the life of Jesus. Indeed, it is a *crux* of Christian theology, but it is not the end of the gospel, atonement extends beyond the cross to the central event: the resurrection. The hope of the gospel is not in death I get to go to heaven (i.e., status), which I sense in many of the

---

[131] For a general survey see Millard Erickson, *Christian Theology*, 3rd ed. (Grand Rapids, MI: Baker Academic, 2013), 713–53; Bruce A. Demarest, *The Cross and Salvation: The Doctrine of Salvation*, Foundations of Evangelical Theology (Wheaton, IL: Crossway Books, 1997).

[132] Irenaeus, *Against Heresies*, V.1.1, *ANF* 1, 526.

[133] Peter Abelard, *Sententie magistri Petri Abelardi (Sententie Hermanni)*, ed. Sandro Buzzetti (Firenze: La Nuova Italia, 1983), 23.

[134] Karl Barth, *Church Dogmatics: The Doctrine of God*, vol. II, part 2 (London: Clark Publishing, 2004), 123; vol. IV/1, 230–35.

atonement theories above. The hope of the gospel of Jesus is the resurrection. The cross is a medium of this transformation, but it is not to be the sole focus of our theological vision. As many womanist theologians will quickly draw out, emphasizing the "goodness" of death and suffering (i.e., praising the cross event) bleeds over into our ethical frameworks about suffering in this life. It can soften the reality that death and suffering are enemies of the vision and hope of God. But not only that, there is a temptation to make God's justice only a post-mortem reality, and that is not something we see in either Jewish or Christian theology.[135] This confusion about God's justice in the here and now, and the overemphasis on the cross as a medium of status renewal, has regrettably become an opiate. It downgrades a vision of hope in the present reality by turning one's eyes toward the future. It is, in some ways, a theological *longtermism*. What implications does this have for our view of violent technology? I suggest it leads to theological impairments.

The cross is not an implement of hope; it is one of shame, death, and suffering. Likewise, death machines are not implements of hope. Yet, as we will see in chapter 3, the Western mind has bought into (literally) an idea that tech can either be a means of deliverance or a means of destruction. Both views distort the reality of what these implements are: the products of socio-political power. When we use technology to violate and force our will upon another being, we undermine the hope of the gospel and equally God's justice. Theologians often talk about God's rectoral justice (cf. Psalm 9:8; 36:6; 50:6; 51:14; 72:1–2) or distributive justice (cf. Deut 32:4; Ps 7:9; 129:4; Isa 26:9; Mal 4:2), and certainly Scripture portrays God as righteous or just. My concern is not how the Bible portrays God as such in the narratives of his works, but how we translate such into our contemporary application of justice here and now. Who is to arbitrate amid the anthropological understanding that all humans are fallen? How will the individual see his orher part in this socio-political struggle for kingdom making and keeping?

In the brutal and bloody episode of the cross we see Jesus embody the command to "love your neighbor" through varies laws (cf. Lev 19:0-35). Instead of saying the cross is violent and necessary, thus, violence is a justifiable reality for Christian theology, the message of Jesus chooses

[135] For an extensive study of this topic see N. T. Wright, *The Resurrection of the Son of God*, Christian Origins and the Question of God (London: Society for Promoting Christian Knowledge, 2003).

a different path. It is wise to remember that Jesus sat with the social pariahs. The woman caught in adultery in John 8 and the women at the well in John 4 are two perfect examples of Jesus listening to the voice of violation and responding in love instead of violence. We are not following Jesus in redemptive violence, but his model of relating to the *other*: women and children during the first-century. Based on the witness of Jesus' ministry, it seems logical to conclude that he would rather us follow him in refusing the victimization and valorization of violence.[136] The history of the female experience is significant enough evidence that traditional atonement theories leave many human persons vulnerable.[137] I wrote the chapters ahead with this question in mind: How do I embrace the antiviolence of the theology of Jesus while also grasping (lamenting) the Western addiction to violent tech?

Conclusion

I ask the reader to consider some strategies formed by Traci C. West, David C. Cramer, Myles Werntz, and Walter Wink. First, we must self-critique our structures and technologies that lead to colonization, racism, ableism, and the subjugation of women.[138] I believe therefore the Bible does not shy away from the bloody and ugly episodes that showcase human nature at its worse. The Bible is a mirror unto the soul, and it invites the reader to see their blemishes. Second, we must resist the narrative that the only options, in response to violence, are either passivity or revolution. Engaging, listening and responding in love to victim-survivors, are ethical commitments we make to our community. When we close our eyes and ears to the pain (past or present) of those that have been *othered* we refuse to give them a *face*. A Christian commitment to resistance means first looking within one's view of violence, what is desirable about X or Y, and why does it have such a strong eschatological force?

---

[136] I am indebted to the work of Carolyn Holderread Heggen's work, *Sexual Abuse in Christian Homes* (Eugene Oregon: Wipf & Stock), 94.

[137] See the survey in David C. Cramer and Myles Werntz in *A Field Guide to Christian Nonviolence: Key Thinkers, Activists, and Movements for the Gospel Peace* (Grand Rapids, MI: Baker Academic, 2022).

[138] Adapted from Traci C. West, *Wounds of the Spirit: Black Women, Violence, and Resistance Ethics* (New York: New York University Press, 1999), 194–99.

Marjorie Hewitt Sochacki reminds us that our assumptions about violence and its implements are based on theological categories.[139] For Christians, these categories include the fall, natural theology, the cross, and ecclesiology. It is not so much that we are rebelling against God in our violence and technology, but against nature and one another's well-being. This is evidenced in the many male theologians noted throughout this chapter. There is a temptation to make ethically ambiguous narratives pristine, and I think this is a result of lacking creativity and a refusal to read such troubling narratives carefully. Womanist theologies, primarily focused on black bodies and experience, offer an alternative vision to violent texts and dare I say technologies. The Church should not justify the destruction of other image bearers or make violent texts pristine. We would do well to share the suspicion of Delores Williams, for example in the narrative of Hagar, by applying a hermeneutic identification ascertainment.[140] Looking intently at troubling texts means examining the traditional ways these scriptures have been read and discerning how they have been employed in our rebellious nature. The cross of Jesus is an important symbol of hope and justice, but as we have seen above, it can be used to justify sacrifice. Yet, as Jacquelyn Grant notes the Bible is a book of hope and a primary source for developing alternative pathways for our desires for revenge and resignation to violent ends.[141]

Womanist theology and ethics also asks us to reject the notion that justice and forgiveness for past and present sin lies on the shoulders of the oppressed. This is why Niebuhr's position is to be rejected. This is exactly the point made by Ida B. Wells in her response to the question of justice and responsibility for lynching in the South.[142] For Wells, and many others, the pathway forward requires reorientation, turning toward something and someone. Many of the womanist theologians orient their hope through the lens of Jesus. I pray we join hands with those like Katie Cannon, Emilie M. Townes, Angela D. Sims, Karen Baker-Fletcher,

---

139 Sochacki, *Fall to Violence*, 161.

140 Delores Williams, *Sisters in the Wilderness: The Challenge of Womanist God-Talk* (Maryknoll, NY: Orbis Books, 1993).

141 Jacquelyn Grant, *White Women's Christ and Black Women's Jesus: Feminist Christology and Womanist Response* (Atlanta: Scholars Press, 1989).

142 Angela D. Sims, *Ethical Complications of Lynching: Ida B. Wells's Interrogation of America Terror* (New York: Palgrave Macmillan, 2010).

Marcia Riggs, A. Elaine Brown Crawford, and Martian Luther King, Jr. to see that anti-violence is a better theological ethic for those who follow Jesus.[143]

In the next chapter, we will examine the roots of our attachments to war and death machines as implements of hope and why we must resist this salvific discourse of benevolence through harmful means. We will also see the power of these machines upon our theological visions and the temptation to remove the *face* of those we deem our "enemies."

[143] See Katie Geneva Cannon, Emilie M. Townes, and Angela D. Sims, eds. *Womanist Theological Ethics: A Reader* (Louisville, KY: Westminster John Know Press, 2011).

CHAPTER 3

DARPA And Death Machines: A Love Story

It is strange, really; without the Defense Advanced Research Projects Agency (DARPA) I would not have the ability to write these words on a digital screen and send them to an editor thousands of miles away through the Internet (developed from ARPANET). DARPA, formerly known as the Advanced Research Projects Agency (ARPA), is the love story between the Department of Defense and the scientific community. Through this bad romance, we get many of the debased technological children that we have grown to love and despise ourselves for loving them. As we saw in the last chapter, Western attachments to war take root in eschatological hope, and not the theological kind. DARPA is the embodiment of this eschatology and understanding its past efforts, present focus, and future hopes will help the reader see the cost of these attachments, whose causation and implementation are steeped in the advancement of the military-industrial complex. First, we will need to pull back the veil on the *who* and *what* that makes DARPA work. Second, it is necessary to understand the political, environmental, and theological costs of this agency. And last, we will consider means of resistance to these visions of advancement for both the Department of Defense (DoD) and civilian life.

The Lingering and Misguided Vision of DARPA

The Department of Defense Secretary, Neil McElroy, under President Eisenhower, issued the DoD Directive 5105.15 on February 7, 1958, officially giving birth to the Advanced Research Projects Agency (ARPA).[144] The question is, what political (or theological) framework led to this vision? The historical context takes place during the launch of Sputnik, which happened a year prior to the launch of ARPA. There were two competing political entities in the inauguration of the space race between the US and the USSR (i.e., the Cold War). What should have been a spectacle for celebration became a vision of doom and fear. The fear that the USSR would have more power than the US or the fear that non-Western governments would use this technology to destroy the American people. This vision led President Dwight Eisenhower to support the development of ARPA and eventually was absolved into the National Advisory Committee for Aeronautics (NACA) which would officially become the National Aeronautics and Space Administration under the Aeronautics and Space Act of 1958.[145] The vision and focus of ARPA in the early years of the 1960s centered around roughly three subjects: space, missile defense, and nuclear weapons. NASA absorbed the space focus early on and took a large portion of ARPA with them, hence the birth of DARPA. The media driven frenzy that led to the inception of ARPA was more about public fear and ignorance than about concerns from President Eisenhower. Little did the public know, at the time, that the President was aware of the science behind the launching of satellites, the launching of ICBMs, and other ballistic payload capabilities, because of the Lockheed SR-71 project. Of course, the public was not aware of this classified research. However, the public outrage or narrative was a blessing in disguise for Lyndon Johnson, the Democratic Senate leader, allowing Johnson a political angle to attack the Republicans. Johnson proclaimed, "soon, they will be dropping bombs on from space like kids dropping rocks onto cars from the freeway

144 "ARPA is Born," *Defense Advanced Research Projects Agency*, www.darpa.mil/about-us/timeline/dod-establishes-arpa, (accessed Nov 17, 2022).

145 See "Public Law 85-567–July 28, 1958," *House.Gov*, 2022, https://uscode.house.gov/statutes/pl/85/568.pdf, (accessed Nov 17, 2022).

overpass."[146] This was politically charged hyperbole, and the President knew it; however, the media coverage had nudged the oval office to respond to public panic.

Missile defense and nuclear weapons became the primary focus of DARPA during the period between 1960 and 1970. Avoiding or preventing technological surprise (cf. Sputnik) were the obsessions of Jack Ruina and J. Licklider. Project AGILE was a major emphasis during this period and some of the outcomes include the adoption of the Armalite AR-15,[147] and the development of X-26B sailplane (cf. Schweizer SGS 2-32), and even Agent Orange. All of which framed the model for the collaboration with civilian researchers for the purposes of providing "improvements in allied nations' capability to meet the threat [of insurgency], and DoD capability to assist them…"[148] As Charles Herzfeld, the Director during this era stated, "AGILE was an abysmal failure, a glorious failure."[149]

Despite the failure of Vietnam, the research model, courtesy of ARPA (now DARPA), captured the eye of the Pentagon and the CIA. Juxtaposed with the political drive of the Cold War, DARPA was recognized for its potential, and this lingers into the present attachment to war-driven research today. Going back to the history of DARPA's innovation and the eventual creation of modern death machines, we pick up in the 1970s-1980s era. Image recognition, natural language processing, computer modeling, and AI were on the radar of DARPA during this period, but the financial wake of Vietnam left little room for this growing appetite. However, research efforts in this era were smaller in pursuit and focused on further "prevention of technological surprise."

While under the presidential umbrella of Richard Nixon's vision to find alternatives to nuclear weapons in responding to the Soviet Union,

---

146 "Recreating ARPA, The Most Successful Research Agency in History," *The Economist*, Jan 23, 2020, https://www.economist.com/britain/2020/01/23/recreating-arpa-the-most-successful-research-agency-in-history, (accessed May 5, 2023).

147 David M. Fortier, "The Adoption of the Black Rifle: Project AGILE: Colt's AR-15 Went from Utter Failure to Adoption in Just Three Years," *Firearms News* 74, no. 21 (2020): 26.

148 See Department of Defense, "Activities of the Research and Development Center – Thailand, Advanced Research Projects Agency," B-167324, December 29, 1971. https://www.gao.gov/assets/b-167324.pdf, , (accessed May 5, 2023).

149 S. Weinberger, *The Imagineers of War: The Untold Story of DARPA: The Pentagon Agency That Changed the World* (New York: Alfred A. Knopf, 2017), 185.

DARPA caught a second wind. Developments and programs in the US military's technology during this era directly correlated to the rise and revolution of personal computing, and even the further development of computer science as we know it today. There is both a focus on common-sense military applications such as command and control, which became of utmost importance during the Cuban Missile Crisis and an unhinged component to the programs of research such as Licklider's cyborg efforts (or man-computer symbiosis).

By the end of the Cold War, it was the "dual-use" technologies, those with applications in both the defense and civilian worlds, that received most of the funding. During this era (1990s-2000s), the Technology Reinvestment Project (TRP) was established to leverage private civilians through corporations and universities to develop technologies under the guise of making life better, but it should be clear to the reader this is a bad romance. It is during this time that programs in biotechnology are being developed and partnerships with the Army are developed to build the Future Combat System, which ends in massive financial failure ($20 billion).

Under Director Anthony Tether in the 2000s, DARPA finally found permanence, at least in the eyes of the Defense Department. Tether became director quickly following the attacks of 9/11. For the first time in decades, DARPA would have proper funding and license to develop technologies that would advance the 'war on terror.' This led to the Total Information Awareness Program, developments in cognitive computing, and further advancement of autonomous systems. Even the youth of America were joining the efforts to help DARPA through the "DARPA Urban and Grand Challenge."[150] These initiatives proved successful and continued through the 2010s as developing technological surprise had finally woven into the fabric of Western societies' idolatry of national security.

All visions are costly. From the earliest visions of ARPA to the present developments happening today through DARPA, the goal has always been to wield power and surprise over *our* enemies. The cost has always been the *soul* of the nation. In the name of national security, it is all too simple to abstract the costs of war-making. The directors, project

[150] The language and vision here are labeled as humanitarian, and the goal is still "dual-usage" technology, www.darpa.mil/program/darpa-robotics-challenge (accessed Nov 16, 2022).

managers, and civilian contractors under the banner of DARPA have successfully created a model of symbiosis between military and civilian life. The question before us is whether this symbiotic relationship is mutually beneficial or rather if one party is a parasite and the other a host?

Through Western eyes, the DARPA model for research and technological development is praised as a "good political design,"[151] but this moral commentary assumes a particular metric for what is "good" and for whom it is "good." The inception of ARPA/DARPA was to be the "imagineers of war." This moniker from Sharon Weinberger's brilliant work *The Imagineers of War: The Untold Story of DARPA, the Pentagon Agency that Changed the World* [152] cuts through the veil of clout. From the inception of ARPA to the political powerhouse that it is today, it has always been about mad science, cost-benefit analysis, and building the American war machine. Director Tether stated the best directors are those that are absorbed by science fiction.[153] Tether, recapitulating the vision of Michael Goldblatt, opened the 2002 DARPAtech conference with this speech:

> Imagine 25 years from now where old guys like me put on a pair of glasses or a helmet and open our eyes. Somewhere there will be a robot that will open its eyes and we will be able to see what the robot sees. We will be able to remotely look down on a cave and think to ourselves, "Let's go down there and kick some butt." And the robots will respond, controlled by our thoughts. It's coming. Imagine a warrior – with the intellect of a human and the immortality of a machine –controlled by our thoughts. [154]

This quote, juxtaposed with the vision and funding history of DARPA, gives witness to the transhumanistic vision of the Department

---

[151] W. B. Bonvillian, "Evolution of U.S. Government Innovation Organization: From the Pipeline Model to the Connected Model, to the Problem of 'Political Design," Presentation at the National Graduate Institute for Policy Studies (GRIPS), GRIPS Innovation, Science, and Technology Seminar, Tokyo, April 2013.

[152] Weinberger, *The Imagineers of War.*

[153] Weinberger, *The Imagineers*, 316.

[154] T. Tether, DARPAtech, 2002 welcoming speech, September 9, 2002 http://archive.darpa.mil/DARPATech2002/presentations/diro_pdf/speeches/TETHER.pdf (accessed May 31, 2022).

of Defense and its love child, ARPA/DARPA. What do I mean by a vision of *transhumanism?* DARPA's flavor runs counter to the rationalistic and humanistic vision of *transhumanism*.[155] It is anti-humanistic because it is a vision that rationalizes suffering, destruction, reducing life to calculations of the existential risks. What contributes to this vision of anthropology and eschatology?

Moving from the theological abstracts in chapter 2, it is helpful to see the direct implementation of DARPA's eschatology present within the mad science of production. DARPA managers have a short shelf-life, only four-to-six-year terms. This half-life does not give managers much time to consider ethical implications, rather it invokes a force of hurriedness. The hurriedness related to the original vision of DARPA is best actualized through creating the technological surprise first.[156] Tether notes, "the term limits are what really make this place. When someone walks in the door here, there's a sense of urgency that you don't find anywhere else in government."[157] This ethos is how an agency with .5 percent of the US defense budget can be extremely efficient. It must be a theological vision that drives them, one that is eschatological: to usher in the eschaton or suffer the wrath of foreign power. But the reader should note that fear is a primary driver and not faith.

If the reader is not convinced, let us go back and consider the words of Eisenhower and his ultimate realization that attaching *hope* to military technological advancement was the only way forward. Consider these words from Eisenhower in 1953:

---

[155] Nick Bostrom is correct in his paper on the development of transhumanistic thought, it is very much a religious vision of human development. He writes, "The quest to transcend our natural confines, however, has long been viewed with ambivalence. On the one hand there is fascination. On the other there is the concept of hubris: that some ambitions are off-limits and will backfire if pursued. The ancient Greeks exhibited this ambivalence in their mythology. Prometheus stole the fire from Zeus and gave it to the humans, thereby permanently improving the human condition. Yet for this act he was severely punished by Zeus. In the myth of Daedalus, the gods are repeatedly challenged, quite successfully, by the clever engineer and artist who uses non-magical means to extend human capabilities. In the end, however, disaster ensues when his son Icarus ignores paternal warnings and flies too close to the sun, causing the wax in his wings to melt," in "A History of Transhumanist Thought," *Journal of Evolution and Technology* 14, no. 1 (2005): 1.

[156] Michael Belfiore, *The Department of Mad Scientists: How DARPA Is Remaking Our World, from the Internet to Artificial Limbs* (New York: Harper Perennial, 2009), 176–77.

[157] Belfiore, *Mad Scientists*, 178.

> The cost of one modern heavy bomber is this: a modern brick school in more than 30 cities. It is two electric power plants, each serving a town of 60,000 population. It is two fine, full equipped hospitals. It is some fifty miles of concrete pavement. We pay for a single fighter plane with a half million bushels of wheat. We pay for a destroyer with new homes that could have housed more than 8,000 people.[158]

The same man in his farewell address in 1961 said the following,

> A vital element in keeping the peace is our military establishment. Our arms must be might, ready for instant action, so that no potential aggressor may be tempted to risk his own destruction…American makers of plowshares could, with time and as required, make swords as well. But now we can no longer risk emergency improvisation of national defense; we have been compelled to create a permanent armaments industry of vast proportions.[159]

What can we credit for this change in posture? Eisenhower, since the launch of Sputnik, had undergone a theological conversion. The question that lingers is whether DARPA is a gospel of truth or temptation? To put it another way, is there empirical evidence that supports that the best pathway for deterrence and national defense is found in standing armies that are technologically evolved? Psychology, philosophy, and theology play a part in answering this question. From the realm of psychology, there is strong evidence that fear can induce aggression, especially in relation to national defense.

The "us or them" operational mode that leads to pre-emptive strikes is nothing new. For example, the Jamestown 1622 massacre was considered a means of justified violence. Whether it is through muskets, drones, or nukes does not really matter. There must be a psychological

---

[158] Dwight Eisenhower, *Address to the American Society of Newspaper Editors*, April 16, 1953, Slater Hotel, Washington, D.C., https://www.presidency.ucsb.edu/documents/address-the-chance-for-peace-delivered-before-the-american-society-newspaper-editors, (accessed Nov 16, 2022).

[159] Dwight D. Eisenhower, "Final TV Talk 1/17/61 (1)," Box 38, Speech Series, Papers of Dwight D. Eisenhower as President, 1953–61, Eisenhower Library, *National Archives and Records Administration*, https://catalog.archives.gov/id/594599, (accessed Nov 16, 2022).

openness to anticipatory aggression. It is not driven by hate or spite, but rather by fear. It may surprise the reader to know that empirical research supports this claim. In 2013, researchers found that when we are faced with a "potential threat" 50% will opt for a preemptive defense, even when there was nothing to personally gain from doing such.[160] DARPA exists today because we want to ensure that we can technologically 'push the big red button' before those we fear might have a chance to do so. Just follow the money. DARPA funding, since 1996–2020 has been on an upward trajectory.

Figure 14. DARPA Funding

obligational authority, in billions of current dollars

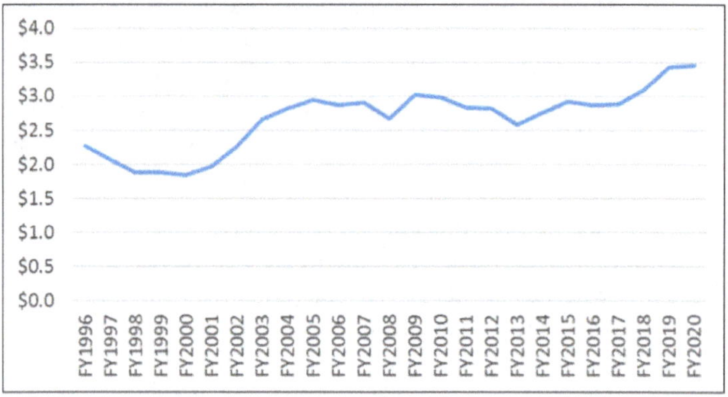

Fig. 2. CRS analysis of data from Department of Defense, Research, Development, Test, and Evaluation Programs (R-1) for FY1998-FY2021.

Eisenhower's address in 1953 was right, the increase of armaments is costly, but there is more to the cost than mere finances. Jean Jacques Rousseau noted in 1782 that conflicts usually start with preventative measures of personal protection but end up as a means to some sort of conquest.[161] This is the telos of all war-making. Technology is simply caught up in this psychological state of fear. The reality is the creation of

[160] Dora Simunovic, Nobuhiro Mifune, Toshio Yamagishi, "Preemptive Strike: An Experimental Study of Fear-Based Aggression," *Journal of Experimental Social Psychology* 49, no. 6 (2013): 1120–1123.

[161] Jean Jacques Rousseau, *A Lasting Peace Through the Federation of Europe*, trans. E. E. Vaughan (London: Constable, 1971), 78–79.

technological implements of violence leads to the creation of more technological implements and further engagement in militarization.

Prior to the events of 9/11, preventive war strategies were considered taboo. But the fear-induced aggression sparked by the falling of the Twin Towers, was too much for the public and Washington to ignore. The preventive war strategy found favor in both Presidents Bush and Obama. Before 2010 only three countries had weaponized drones, but as of 2020 over 94 countries have developed or acquired them.[162] The United States has led the charge in the developing machines of death; the most 'successful' appears to be the atomic bomb under the Manhattan Project. Yet, we must look beyond or behind these implements for the moment to see the entanglements of the theology and science.

This reality goes back to the etymology of the word "robot," which stems from the Czech *robota* ("forced labor"). Whether the artificial object is conscious is of no significance at present.[163] What matters most here is the enframing, that slaves always serve a master. DARPA did not create this epistemological problem, but they certainly have profited from it. In essence, this is what Yarden Katz labels as *artificial whiteness*. Defense projects are funded to serve an ideology and political agenda, even when they are under the guise of making the world a 'better' or 'safer' place. Katz's work in *Artificial Whiteness* is a witness to how AI (like other past implements) is about *whiteness*. According to Katz, *whiteness* is maintaining power. Not only is this process about holding onto a political position but it is a means of oppression as well.[164]

The major narrative about drone warfare is that it is *clean* or *precise* and that it saves the lives of our service men and women (i.e., Ronald Arkin).[165] In this reading of AI systems and war-machines, the entity comes into existence *tabula rasa* ("blank slate"). Claiming that AI and death machines come from a view of nowhere is political forgery. This

---

162 Seth J. Frantzman, "As Drones Proliferate, Authoritarian Regimes Profit," *Newlines Institute*, June 23, 2021, https://newlinesinstitute.org/technology/as-drones-proliferate-authoritarian-regimes-profit/, (accessed Nov 16, 2022).

163 This subject matter is deeply important to me, and I believe we should not make slaves. Here I simply mean we cannot unpack it carefully within this chapter. We will return to this more in the section addressing the *face* of death machines.

164 Yarden Katz, *Artificial Whiteness: Politics and Ideology in Artificial Intelligence* (New York: Columbia University Press, 2020), 16.

165 This is a major thesis of researcher Ronald Arkin in *Governing Lethal Behavior in Autonomous Robots* (New York: Routledge, 2009).

veneer is a major part of the rebranding of AI as a "social good" and the DARPA research that funded elite universities like Harvard, MIT, Carnegie Mellon, and Stanford.[166] Some of these DARPA projects include the following:

> 1985–1987 The Autonomous Land Vehicle (ALV)
> 1993–1997 The Unmanned Ground Vehicle (UGV)
> 1998–2000 The Tactical Mobile Robotic Program
> 1999–2003 The Mobile Autonomous Robotic Software Program
> 2001–2002 The Unmanned Ground Combat Vehicle Program
> 2001–2003 The Future Combat Systems (FCS) Communications Systems, Integration, and Demonstration Program
> 2002–2005 The Mobile Autonomous Robotics Software Vision

Guess which leading roboticist has been a part of all these programs? Ronald Arkin, the same voice telling the public that these systems make war *better.*We have already seen that the mere research and creation of death machines leads to the creation of more entities, not less. But the promises and false hope do not end there.

In order to continue the creation of death machines that will give technological advantages on the battlefield the public has to see them from a particular frame. Specifically, the public must see the human right of our enemies as less than the human rights of our citizens. For example, state sanctioned or sponsored violence is acceptable (i.e., Abu Ghraib) and in the face of the threat of terrorism, the posture shifts from violence to survivability. Thus, to protect Western borders, it is justifiable to violate other sovereignties and their borders.[167] Yet, this ideology goes beyond mere nationalism. It is now a promise of biomedical healing and enhancement–through war we can heal the wounds of war. Here is where the Department of Defense and DARPA capitalizes further on their victims: veterans of international war.

Bionic prosthetics emerged and expanded during the invasions of Afghanistan and Iraq. Advancements in prosthetics, as Erin O'Conner notes was a "symbol of all that was possible in the modern world of manufacture, a walking advertisement for the personal and social

[166] Katz, *Artificial Whiteness*, 66–89.
[167] Judith Butler, *Frames of War,* 47–48.

benefits to be had from a full-scale embrace of machine culture."[168] Before getting back into our critique of the DARPA machine, let us explore why this anthropological phenomenon emerges. In the first chapter recall that we briefly explored the concepts of sin and fall. In my doctoral research on robotic futurism, I noticed a dearth of the concept of depravity in varying views of futurism.[169] To put it in simple theological terminology, there is no concept of a sin nature.

Futurism's Metaphysic and Death Machines

The great minds of roboticists like Hans Moravec, Marvin Minsky, Ray Kurzweil and Rodney Brooks give further insight into this theological dearth. For Kurzweil, the technological vision is ultimately about radical life extension and hopefully the *uploading* of the mind to a new permanent substrate.[170] Undergirding this view is a neo-Cartesian perspective of materialistic patternism: which sees the foundation of all reality as mere patterns of information.[171] The telos of life in this anthropology is data transfer. A person is a combination of informational bits stored in physical hardware (i.e., neural network). This is made possible by holding a view of technological evolution, which is oriented toward the development of knowledge. A primary assumption here is Kurzweil's "Law of Accelerating Returns" argument which states there is an "inherent acceleration in the rate of evolution, with technological evolution as a continuation of biological evolution."[172] This assumes that the natural process of evolution, which is inherently chaotic, leads to the ordered and exponential development of information, with humans

168 Erin O'Conner, "Fraction of Men": Engendering Amputation in Victorian Culture," *Comparative Studies in Society and History* 39, no. 4 (1997): 742–777, 767.
169 Joshua K. Smith, *Robotic Persons: Our Future with Social Robots* (Indianapolis: IN, Westbow Press, 2021).
170 Ray Kurzweil, *The Age of Spiritual Machines: When Computers Exceed Human Intelligence* (New York: Viking, 1999), *The Singularity is Near: When Humans Transcend Biology* (New York: Viking, 2005), *The 10% Solution for a Healthy Life: How to Reduce Fat in Your Diet and Eliminate Virtually All Risk of Heart Disease and Cancer* (New York: Three Rivers Press, 1994); Ray Kurzweil and Terry Grossman, *Fantastic Voyage: Live Long Enough to Live Forever* (New York: Plume, 2005). Ray Kurzweil and Terry Grossman, *Transcend: Nine Steps to Living Well Forever* (New York: Rodale, 2010).
171 Kurzweil, *The Singularity is Near*, 5.
172 Kurzweil, *The Singularity is Near*, 7.

being the apex of biological evolution. Indeed, but what keeps humans from reaching the pinnacle of singularity?

What ails this world, according to Kurzweil's anthropology, is ignorance. The social and political problems that face this world, including violence, are reduced to the technological. Emerging technology, in this vision, will solve the global problems of poverty, warfare, and inequality. Hans Moravec shared Kurzweil's vision.

Similar to Kurzweil, Moravec's anthropology envisions humans as reducible qualia. Again, there is subscription to body-mind dualism or the Cartesian flavor in this thinking, in particular that the mind is a material property-thing. However, the dualism is slightly different because Moravec and Kurzweil see the human person as pattern-identity and not body-identity.[173] It is uncertain from this metaphysic how identity will be preserved through change. This vision, exponential evolution toward the singularity, where emergent technology is the causal agent, frames the acceptable cost of violent technology. After all, it is a mere blip on the radar of the journey toward *postbiology*. What a convenient view to prop up one's desire for colonialism.[174] Before further critiquing this confused metaphysic and how it has led to many current developments through DARPA funding, let us consider one more futurist that is also steeped in the death machine ethos, Marvin Minsky.

Emeritus professor at MIT, Marvin Minsky, has also contributed to the *theological* developments that correlate to the development of modern death machines. Minsky's research as evidenced in his works *The Emotion Machine* and *Society of the Mind* places much emphasis on the mind as the

---

[173] Han Moravec, *Mind Children: The Future of Robot and Human Intelligence* (Cambridge, MA: Harvard, 1998), 116.

[174] In case the reader believes this is hyperbolic, consider the words of Moravec in *Robot* comparing today's humans to a tribe in New Guinea, "The naked inhabitants, some with stone spears, were driven into paroxysms of confusion and religious fear and awe by the giant roaring silver birds that alighted near their mud-thatch villages to release droopy-skinned white men without genitals who, among too many wonders, captured their souls in small black boxes labeled Kodak," 3.
many wonders, captured their souls in small black boxes labeled Kodak," in Moravec, *Robot: Mere Machine to Transcendent Mind* (New York: Oxford, 1999), 3.

loci of the person.[175] Also, following Descartes' mechanistic view, Minsky explains the functioning of the mind in computational terms. Humans are a type of machine, running on biological or natural mechanisms.[176] If you were wondering, yes, there is a deterministic framework in this perspective as well. Minsky writes, "everything, including that which happens in our minds depends on these and only these: A set of fixed, deterministic laws. A purely random set of accidents."[177] Unlike the former roboticist above, Minsky does highlight emotion as a significant part of human intelligence. Axiological assumptions, experiences, and perceived values are important to the development of the mind. Unfortunately, everything is reduced to physical components, and again does not account for the common philosophical issues of identity through change. For example, consider how these thinkers might not be able to solve the simple challenge of the identity of Indiscernibles principle: $x=y \rightarrow \forall F(Fx \leftrightarrow Fy)$.[178] Let us explore further why this matters for creating death machines and for metaphysics in general.

I do not wish to chide the roboticists above. In fact, much of what they formulate in their metaphysics is *common sense* thinking in modern biological and physical science. Of course, this does not mean that it is factual, but an admission that it is a warranted and justified view, in their opinion. Again, there is a century-long struggle with Aristotelian notions of actuality and potency, granted some notions need to be challenged in a modern era that values equality. In fact, I believe there are numerous things that Descartes's dualism and mechanistic view gets right. Even Richard Rorty acknowledges that modern physics is somewhat of a distorted (i.e., funhouse) mirror of reality. All that to say, we need a sense of critical realism in between our struggle with empirical science and the philosophy of nature–no one gets it exactly right.

---

175 Marvin Minsky, *The Society of Mind*, (New York: Simon and Schuster, 1986); *The Emotion Machine: Commonsense Thinking, Artificial Intelligence, and the Future of the Human Mind* (New York: Simon & Schuster, 2007).

176 Minsky, *Society of Mind*, 41.

177 Minsky, *Society of Mind*, 306, emphasis removed.

178 Originally formulated by Wilhelm Gottfried Leibniz in *Discourse on Metaphysics and the Monadology,* Dover Philosophical Classics, trans. George R. Montgomery (Mineola, NY: Dover, 2005), xi–53.

What is clear, or at least I have tried to make it so, is that physicalism undergirds the metaphysics of robotic futurism. It is a real problem, as Noam Chomsky observed in 1988, that notions about physics become notions about the physical world and its bodies. It is circular reasoning (cf. defining mass as a quantity of matter).[179] Indeed, but still one of the major problems that robotic futurism leaves in its wake is a proper account of causality. I understand that this leaves my Aristotelian sympathies exposed. Be that as it may, it still matters deeply for the assumptions being made in the realm of quantum mechanics. Is there potential for the movement from A to B to be actualized? This is a question that many in the field of robotics assume without proper metaphysical warranting. In Aristotelian physics, objects in motion are actualizing potential, but not in a complete way. The object is not actualized until the motion is complete.[180] Determinism is not entailed by causality,[181] which seems to be a major assumption in the DARPA doctrine. Regrettably, the rejection of Aristotle leaves the physics of computation in metaphysical peril.

What are the differences between the state of computer vision and that of human vision? Or what of the state of information of a disk and of the human brain? In a technical sense (i.e., syntactic), a *bit* is a basic unit of information consisting of two possible values, which are represented as 1 or 0. Listen to the words of philosopher David Chalmers on this when he writes:

> Each "bit" on the compact disk has an independent effect on the compact disk player, so that each location on the disk can be seen to realize a two-state subspace of its own. Putting all these independent effects together, we find a combinatorial structure in the space of total effects of a compact disk, and so we can find the same combinatorial

[179] Noam Chomsky, *Language and Problems of Knowledge* (Massachusetts, MA: MIT Press, 1988), 144.
[180] Aristotle, "Physics," bk. III.
[181] Cf. Phil Dowe on radioactive decay, in *Physical Causation*, Cambridge Studies in Probability, Induction, and Decision Theory (Cambridge: Cambridge University Press, 2000), 22–23.

structure in the information space that the compact disk realizes.[182]

Computers are processors of syntactic information. Electronic states provide the causal path for the information that is embodied and generated through a particular media (i.e., screen). When I insert a flash drive into a USB-A port on my computer, it will recognize this device and will run a script of instructions to allow me to access the information on said drive. Although it is not the one-to-one transfer of semantic information, computation is still the transfer of embodied intelligence. But this still does not tell us what the potential features these media can take on, rather, it tells us the effects. The uncertainty has led many to believe there is no intentionality in this world of mechanistic computation. Alex Rosenberg, Peter Godfrey-Smith, and Richard Dawkins, see the world and evolution as a type of algorithm.[183]

Concerning the mind as computer argument, assumed above, Paul Churchland has further advanced this thesis in computation neuroscience. Roboticists did not create the metaphysical problems they employ. Rather, we can place that blame more directly on the philosophy of Churchland during his 1981 paper, "Eliminative Materialism and the Propositional Attitudes." Essentially, neuroscience will eventually explain and prove that the conception of psychological phenomena is false. There are no immaterial states for Churchland.[184] Okay, but what does this say about common-sense views about mental states? It says nothing. Churchland appeals to a future when we "construct a new system of verbal communication entirely distinct from natural language."[185] So at some point Churchland, using natural language at present, will devoid his very statements when his new theory comes to light, in which the former propositions will be made void or ironically

---

[182] David Chalmers, *The Conscious Mind: In Search of Fundamental Theory* (Oxford: Oxford University Press, 1996), 282.

[183] Richard Dawkins, *The Blind Watchmaker: Why the Evidence of Evolution Reveals a Universe without Design* (New York: W. W. Norton & Company, 1986), 111; Peter Godfrey-Smith, *Other Minds: The Octopus, the Sean, and the Deep Origins of Consciousness* (New York: Farrar, Straus, and Giroux, 2016), 111–12; Alex Rosenberg, *Atheist's Guide*, 99–100.

[184] Paul Churchland, "Eliminative Materialism and the Propositional Attitude," *Journal of Philosophy* 78, no. 2 (1981): 1.

[185] Churchland, "Eliminative Materialism," 19.

eliminated. There are no such things as beliefs, desires, and so on, then what makes up Churchland's perceptions about this theory?

Perhaps our friend John Searle is of help in discerning a pathway forward in diagnosing this religious devotion to future progress. It runs similar to the views of logical positivism in the twentieth century, in which there was very much a dogmatic commitment to empiricism. However, one of the reasons this perspective fails is because of its exclusion of the data on human experience. Searle said in *the Rediscovery of the Mind*, "If you are tempted to functionalism, I believe you do not need refutation, you need help."[186] This is where we are presently with DARPA and with the roboticists that make death machines a reality. Although I disagree with Searle's pessimism about how one can know whether mental states are physical states, at least he does acknowledge the causal relationship between the two. The pathway toward healing is to first look at the human experience (i.e., mental states), not disregard it. Furthermore, through examining the human creator's desires, emotions, and hopes we can reveal a great deal about the telos of their machines. I understand that within Darwinian mechanics teleology is considered an imposition of the observer (e.g., Searle's argument).[187] However, it is a grave mistake to discount the Aristotelian distinction between *essence* and *proper accidents.* How does this apply to death machines? Just because roboticists and DARPA do not include final cause within their *reductio* arguments for death machines does not make it factually true. These machines have a telos and a formal cause, and to give a hint, that causal power is not evolution. This reality gets lost in the reductionism present throughout the DARPA doctrine. Or perhaps it is a necessary condition for the ultimate biological growth of the human person into the *posthuman.*

Death Machines and the Fall

In the absence of a religious perspective or framing of anthropology, political ones take its place. The Augustinian vision of human nature,[188]

---

[186] John Searle, *The Rediscovery of the Mind* (Boston, MA: Bradford Books, 1992), 9.

[187] Searle, *Rediscovery of the Mind*, 51–52, 237–38.

[188] Jesse Couenhavon notes that "Augustine is known as *the* great theologian of sin." In "Augustine," in *T&T Clark Companion to the Doctrine of Sin*, eds. Keith L. Johnson and David Lauber (New York: Bloomsbury T&T Clark, 2016), 181, emphasis original.

one that views humans as totally depraved and inherently sinful was replaced during the 1700s with the mechanistic view of human nature. The religious framing was not replaced or substituted overnight. It was a gradual and constant shift from Descartes, Bacon, Hobbes, Kant, Fichte, and Feuerbach. That is not to say this is a new phenomenon, but there were new media and significant developments in this philosophy that made it successful. One's view of anthropology is the pretension to one's application of technology. When there is no fall or *brokenness* in one's anthropology there are no limitations on metaphysical, teleological, and ethical boundaries.

Another challenge to the current metaphysics of DARPA and the consensus in AI ethics is the problem of evil. The Darwinian account of anthropology does not account for telos, as noted above, but also for the suffering found in nature. Theists typically accept two types of evil: moral and natural. Moral evil involves the exercising of freewill, even if that will is limited in scope, of a creature that can be attributed blame for an action or thought. Natural evil is different, for there is no blameworthy creature acting with freewill. It is logical to place much of animal suffering in the natural evil category (i.e., disease, injury, predation). There exists within classical theism a *curse* on nonmoral natural order. It stems from the episode in Genesis 3 where the serpent, man, and woman rebel against God through creation. God curses the ground.

> Cursed is the ground because of you;
> Through painful toil you will eat of it
> All your days (Genesis 3:17)

The moral evil is shared by all parties involved, not just the human Adam. The curse is also directed toward the Serpent:

> Cursed are you above all livestock,
> And all the wild animals!
> You will crawl on your belly,
> And you will eat dust,
> All the days of your life (Gen. 3:14)

A similar sentiment is communicated through the words of Noah in Genesis 5:29 where he recalls the painful toil and cursing of the ground. The apostle Paul echoes this view in Romans 8:19–22, "For we know the whole creation groans and suffers the pains of childbirth together until now." While some biblical scholars might object, I agree with Richard Hays that Paul is using Edenic language in the book of Romans. Thus, all suffering, and evil is a result of what the Christian theologian would label as the "fall." However, there is a temptation to see the fall and pain/suffering in general as objections[189] to *causa dei* (CD), there are at least four:

- Pre-Adamic Pain/Suffer: how are PAP and the fall related? What is the moral good that comes from the permission of PAP and what is the necessary condition to acquire it?
- Fragility: Why does God create the cosmos in such a fragile state, knowing that the condition of its *goodness* depends on a pair of child-like humans?
- Paradisiacal Motivation: If Adam and Eve are in paradise, why do they desire something outside or foreign to their utopian state? What motivation is there for such desire within rational and free beings?
- Value of Freedom: How does the value of freedom outweigh the prevention of pain and suffering? Granted, we all value freedom, but should it come at the risk of causing cosmic pain and suffering?

While we do not have the time or space to delve into these objections, I mention them because even in light of agnostic and atheistic views of the cosmos the problem of animal suffering asks the question; why does it exist? But there is also another reason why animal suffering is important to the AI and robot ethics dialogue. It was once a standard to view animals as machines. The Cartesian view held that animals are automata and unconscious. This means that there is no continuity to their pain. Now this does not mean that Descartes believed animals did not feel pain, in fact, in his *Meditations* we see the inverse of such. Animals, according to Descartes, through either their brain or sensory neural pathways, have memory and sensation (e.g., pain).[190] Mostly, modern ethology confirms Descartes' presumptions with variations.

---

[189] For an in-depth analysis and critique see, Michael Murry, *Nature Red in Tooth & Claw: Theism and the Problem of Animal Suffering* (Oxford: Oxford University Press, 2011), 73–106.
[190] René Descartes, *The Philosophical Writings of Descartes*, I–III, ed. John Cottingham, Robert Stoothoff, Dugald Murdoch, and Anthony Kenny (Cambridge: Cambridge University Press, 1991), 54–5, 303, 366.

In ethology, there are two flavors: Weak Cognitive Ethology and Strong Cognitive Ethology. The weak view holds that there are mental states present in the animal. Observation of the animal's behavior and cognitive abilities is one thing, but observation of mental states is another. The strong view follows similar empirical methods of observation but goes further to attribute mental states.[191] Again, we are left frustrated because we ask of empiricism what only metaphysics can offer. We have no means of observing one's second order awareness through behavior. Ethology aside, this shows the reader the relevance of axiological assumptions within the realm of rational thinking. The mental states that are not isomorphic to physical states (whether they are dependent upon the latter or not) have causal power over the will. No one is immune, no not even one. Here I reveal my cards, so to speak, I do believe in a fall and that relates to my thinking about total depravity: all of humankind is broken. Not merely humankind, but as we see, in brief, that all creation is ruptured, even at the sub-atomic level. As I mentioned in chapter one, we must be careful with fatalistic framings of the fall and sin nature.[192] It matters what ideology is enframed within the creation and implementation of our machines.

Are death machines *fallen*? Yes, death, suffering, and toil are part of the curse of God upon humanity. I see war and its implements as very much a part of human depravity. Much like the emergence of personhood, the context of the character within the narrative matters. When dealing with robotics research that is meant to be fluid between national defense and consumer markets, it is difficult to simply access the implement as neutral or sinful if we are speaking theologically. Violence always needs implements as a means to power.

The more society accepts political violence as a means of security and health, the more we are able to justify violence in the world. Listen to the words of our friend Hannah Arendt,

> The organic metaphors with which our entire present discussion of these matters, especially of the riots, is permeated–the notion of a 'sick society,' of which riots are

---

191 David DeGrazia, *Taking Animals Seriously: Mental Life and Moral Status* (Cambridge: Cambridge University Press, 1996).

192 I will go deeper into the modern psychology of this framework as well in the next chapter.

> symptoms, as fever is a symptom of a disease–can only promote violence in the end. Thus, debate between those who propose violent means to restore 'law and order' and those who propose nonviolent reforms begins to sound ominously like a discussion between two physicians who debate the relative advantages of surgical as opposed to medical treatment of their patient. The sicker the patient is supposed to be the more likely that the surgeon will have the last word.[193]

It is tempting to oppose the implement of war as a violent object, but we must remember it did not create itself and force its will upon humanity. Humanity is supposed to be evolved, sophisticated, and diplomatic, yet here we are in the 21st century spending billions of dollars on national security while children die by the millions from preventable causes, mostly pneumonia and diarrhea.[194] The creation and implementation of advanced weapon systems are not the cure for what ails human nature. Yes, we are a "sick society," but it does not fate us to war-making. In fact, to continue the trajectory we are on now, we are doomed to self-destructive violence driven by greed and fear.

The fall to violence and the creation of its implements has much to do with human ontology and the doctrine of hamartiology (i.e., sin). An intrinsic part of human anatomy is aggression, or to put in theological terms, rebellion.

While I do not subscribe to macro-evolutionary theory, I know that many in the scientific community prefer a theory of the soul and personhood to (at least) account for such. Like all heuristics, there must be axiological assumptions, and this is equally true for any theosis or *union with God* (i.e., deification). Even if one does not accept this axiom, that the ultimate good is union with God, one still has a sense that the world is not what it ought to be. Thus, some explanation of the problem of evil must be accounted for by all those who desire to describe the *good* of life. Even in evolutionary theory there is a wrestling with the violence intrinsic to nature.

---

[193] Hannah Arendt, *On Violence*, 74.

[194] "Child Mortality and Causes of Death," *WHO*, Child mortality and causes of death (who.int), www.who.int/data/gho/data/themes/topics/topic-details/GHO/child-mortality-and-causes-of-death, (accessed Nov 16, 2022).

Irenaeus saw humans as free creatures that, even in the garden, were not born ready-made for deification. Rather, free creatures are made with the potential and capacity, through suffering, pain, and redemption, to grow into the likeness of God. The greatest good in the world, union with God, is possible through a reality that allows for the most hideous suffering. This is not the hard determinism of Augustine, which makes the most appalling suffering both *willed* and *directed* by God, but rather a metaphysical reality that must be present for the process of soul making to exist. It is the potential for suffering that provides the necessary condition for the creature to develop the highest of virtues. Numerous critics of Irenaeus point out that animal suffering seems to provide some defeaters to this view, primarily that it cannot be formulated for the problem of animal suffering.[195] I do not think that Irenaeus states that all suffering is deserved, but rather that suffering is a part of the development of virtue if it is framed in a theodicy. Paul writes something similar in Romans, "And not only this, but we also exult in our tribulations, knowing that tribulation brings about perseverance; and perseverance, proven character; and proven character, hope; and hope does not disappoint, because the love of God has been poured out within our hearts through the Holy Spirit who was given to us" (NASB, Romans 5:3–5). If Irenaeus is right about the conditions of soul-making, then being made in the *image* is simply a precondition for being made in the *likeness of God.* In this metaphysics, the potential is there from creation, but the actualization is not a reality. Irenaeus believed that our sinful nature and childishness keep us from ever truly being made into a soul that is unified with God, there must be a formal cause and that is Jesus the redeemer. In *Irenæus against Heresies*, he writes about this process,

> Since the Lord thus has redeemed us through His own blood, giving His soul for our souls, and His flesh for our flesh, and has also poured out the Spirit of the Father for the union and communion of God and man, imparting indeed God to men

[195] Nancey Murphy, "Science and the Problem of Evil: Suffering as a By-Product of a Finely Tuned Cosmos," *Physics and Cosmology: Scientific Perspectives on the Problem of Evil in Nature*, eds. Nancey Murphy, Robert John Russell, and William R. Stoeger, Vatican City: Vatican Observatory (Berkeley, CA: Center for Theology and the Natural Sciences, 2007), 131–52. Cf. Trent Dougherty. *The Problem of Animal Pain* (London, UK: Palgrave Macmillan, 2014).

> by means of the Spirit, and, on the other hand, attaching man to God by His own incarnation, and bestowing upon us at His coming immortality durably and truly, by means of communion with God,–all the doctrines of the heretics fall to ruin[196]

How does this view fit into modern views of anthropology and "fall" theories? It shows that the predicament that human nature is more complex than post-Irenaeian views of the fall. One component that fits square into the discussion of DARPA above is that self-centeredness predates the fall. While our original father and mother had the capacity for union with God, there was never the capability for them to achieve this state without the intervention of God himself. Christoph Wassermann traces this evolutionary process where humans struggle to master their environment: Humans move from Pliocene to Pleistocene eras. The vegetarian lifestyle did not survive drought and the violence of ecological change. The death of animals, and the invention of tools through metallurgy were necessary conditions for the survival against nature and other hominids.[197] Basically, human survival entails violence. While I do not believe we have to go back to the fossil record here to validate this claim, I can appreciate what is being said here in both Wassermann and Irenaeus, within human history, there is a precondition and physiological biases toward aggression and rebellion. I think this is much softer than the determinism of Augustine, but not by much. Even more troubling is the evidence found in neuroscience.

Adrian Raine's *The Anatomy of Violence*[198] reinforces the idea being communicated through Irenaeus (cf. John Hick), Wassermann, and Schleiermacher, that there is great reason to believe that our pre-fall condition was not pristine. In this book, Raine seeks to determine whether there is a genetic predisposition to violence. If the evolutionary theory is true, then we used aggression to develop and become the modern humans we are today, advanced, civilized, and fill in any other

---

196 Irenaeus of Lyons, "Irenæus against Heresies," in *ANF* I, eds. Alexander Roberts, James Donaldson, and A. Cleveland Coxe, vol. 1, The Ante-Nicene Fathers (Buffalo, NY: Christian Literature Company, 1885), 527.

197 C. Wassermann, *Struktur und Ereignis: inter disziplinäre Studien in Physik, Theologie und Philosophie* (Faculté-autonome de theologie protestante, 1991).

198 Adrian Raine, *The Anatomy of Violence: The Biological Roots of Crime* (New York: Penguin, 2013).

number of adjectives to meet your fancy. But Raine's work starts with examining why modern parents kill their infants; so, what is happening, are we evolved or still in the process of evolving? The harsh reality of life is that we are surrounded by violence, and while, yes some are genetically predisposed toward such we also have attachments and derive pleasure from it.

Since the scandalous findings of Gasare Lombroso (c. 1871), the study of the inner workings of how the brain and the human environment have helped criminology understand violence. Our unique genetic makeup and personality influences the decisions we make. Lombroso's horrible philosophy of race aside, he opened the door for the field of neurocriminology to exist. Essentially, the question at the turn of the century was whether some were genetically predetermined toward aggression and violent behaviors. We will get more into the psychology of this in the next chapter, but for now, I introduce this for a couple of reasons.

First, violence and the implements of violence are nebulous constructs that cannot be simply reduced to either a religious framework or some biological characteristic. As Raine's work will show in the next chapter, it is a combination of environments and genetics. Conditioning is also a major component of this process. Millions of people play violent video games, but 99% of the time, it does not translate to aggression in the real world, unless we count thrown controllers and vulgar language. More to this, violent sports and films occupy a critical part in the entertainment industry, but does this does not lead us to the conclusion that the millions of people enjoying this media are psychopaths or inherently violent?

Second, the acceptance of political violence forces us to come to terms with something about human nature. Aggression and violence are a part of us. For the longest time, I have heard theologians claim that pride is the root of all sin. Satan's fall from heaven, as well as Adam and Eve's desire to experience the knowledge of the "tree," seems to indicate that pride is, in fact, the foundation of our fallenness. However, the first begotten human man (remember, Adam was made) was Cain. His desires led him to violence. Human emotions have both benefits and impairments. Emotions live within a societal bond, as does all of life. This understanding is paramount because it helps the reader understand the social sin of creating death machines, these implements break societal bonds. Why is this so important?

With both DARPA and the DoD, in the West, there must be an anthropology that permits the destruction of one human over another. This is by no means unique to death machines. In fact, this aggression is prevalent in theological communities as well. Countless men, women, and children have been killed "justly" in the name of a god. My hope is that through awareness of this sin nature, humanity might not repeat this cycle. The implementation of these technologies is simply the elite *nudging* the public in the right direction toward evolution. In 2022, there was a major controversy about an AI chatbot, LaMDA, whom Google engineer Blake Lemoine claimed to be a *person*.[199] The nature of responses to this phenomenon has been tragically predictable. Despite not having yet solved the problem of other minds or being able to answer the question, "what is it like to be an AI model," scholars abound in claiming that LaMDA is not worthy of moral consideration based on capacities and properties. What is fascinating about this response is that it bears witness to the discussion above about the metaphysical assumptions that drive technological innovation, in particular the vision about evolution as a theological theme of progress.

One problem (of many) with the views of robotic futurism above is that it propagates the very thing which it claims to prevent: violence. DARPA has successfully created a symbiotic relationship between public research and war-making. The rationale is that through war-making we can eliminate the threat of war, but in order to do this we must have the best gizmos and gadgets. Evolutionary assumptions aside about human development and progress, even a quick glance into the history of World War I gives evidence that the "Great War" only made room for World War II. As DARPA knows, the elimination of war is never the goal here, but rather the goal is, and has been since AGILE, technological superiority of our enemies. This logic about technological salvation makes a promise it can never keep because violent implements only lead to more violence. Juxtaposed to leading to more violence globally, this worldview is politically naïve. Technology cannot solve the problems that it creates, neither can humans ultimately. Not to mention the

[199] Carissa Véliz, "LaMDA is Nothing Like a Person: This is Why," *The Wire*, June 22, 2022, https://science.thewire.in/the-sciences/why-lamda-nothing-like-a-person/, (accessed Nov 16, 2022).

continuously unraveling of ethical problems that are exhibited in AI systems like sexism, racism, and ethnocentrism. The point of futurism's theology is this: the elimination of the body and its fallenness. Vicki Kirby reminds us of this truth in her critique of neo-Cartesianism, because it makes the biopolitics of the technophile easier to operationalize. Kirby is right, the body will never be obsolete, nor will the affective textures of the mind.[200]

On another front, the *telos* that DARPA and its scientists attach to their views of evolutionary theory are highly questionable. It is a means of replacing theological *telos* with evolutionary ones. What I mean by this is that under the Enlightenment dogma of *human progress*, futurists attach values and meaning to the process of evolution. At best this attachment calls into question the reliability of judging whether those values or worthy of being labeled *good* or simply as a means of survival. Stephen Gould, Philippe Verdoux, and Antje Jackelen have also made this critique of this view of evolution.[201] Such views of determinism and myopic visions of human progress are not only poor science but also serve as an insurance policy against losing power at the bodily cost of our fellow humans. It fates a particular technological trajectory (cf. Kurzweil and Moravec especially), one of quietism. If violent technologies are merely following a *natural* pathway there is no point in trying to change it. The present problems of human history will (somehow magically) dissolve as more technology develops.

Perhaps there is a temptation to blame the object or the implement of violence. I do not believe that death machines are inherently sinful, that they as artefacts somehow inherit the original sin of humanity. However, all technology, because it comes from the *cursed* ground, is corrupted. Evidence can be seen in the reality that there are glitches in the code as both a result of the *curse* of the ground and the biased nature of the coder. This is not a moral commentary of the nature of the coder

200 Kirby writes, "…cybernauts tend to rejoice uncritically in their Cartesian inheritance that would regard the body as 'obsolete, as soon as consciousness itself can be uploaded into the network," in *Telling Flesh: The Substance of the Corporeal* (New York: Routledge, 1997), 138.

201 Stephen Jay Gould, *Full House: The Spread of Excellence from Plato to Darwin* (New York: Harmony Books, 1996), 168-169. Philippe Verdoux, "Transhumanism, Progress, and the Future," *Journal of Evolution and Technology* 20, no. 2 (2009): 50; Antje Jackelen, "The Image of God as Techno Sapiens," *Zygon* 37, no. 2 (2002): 296.

per se, but a witness to my view of anthropology. As a pastor I know all people can do *good* regardless of confession or creed (or lack thereof). I also know that people act depraved, even when they do not understand the desire or decision that was acted out. Lest the reader thinksthat this is simply pessimistic projections from experience, many sociologists bear witness to the brokenness of humanity (or human evil, vice, and corruption).[202] It is not merely that we have not evolved enough or that we are ignorant, but as neuroscience tells us, we are inherently aggressive about self-preservation as emergent persons. Where does this all leave us?

There are a few theological conclusions that should be made. First, violent technologies or death machines are moral actors of indirect actions, and this is why they hold extrinsic value in the eyes of the state. Movements that desire to ban killer robots sometimes fail to recognize this truth. Prohibition against the object is not enough; there must be consideration of *why* these implements are valued by political entities. The question is whether they should have intrinsic or extrinsic value in the eyes of the public, or are they merely tools to be demonized?

Much of the current debate about personhood and moral consideration of robots is situated around the murky concept of "consciousness" unfortunately. Granted, there is something to this argument, but unless we have solved the problem of other minds it is doubtful that we will move beyond the impasse toward fruitful deliberations (cf. the LaMDA discussion). Rather, I believe, like David Gunkel, Josh Gellers, Kate Darling, Jacob Turner, and Sven Nyholm, that there can be great societal value in giving the robot a *face*. Why? Whether the artefact is self-aware, feels pain or emotion, has desires, and so forth has little to do with the human experience of an entity. There can be no room for doubt that humans anthropomorphize non-human objects, even when there is no question about the qualia of the entity.

---

[202] Cf. Roy Baumeister, *Evil: Inside Human Violence and Cruelty* (New York: W.H. Freeman, 1999); Arthur Miller, *The Social Psychology of Good and Evil* (New York: Guilford Press, 2005); Michael Shermer, *The Science of Good and Evil* (New York: Owl Books, 2004); Margaret Breen and Richard Hamilton, *This Thing of Darkness: Perspectives on Evil and Human Wickedness* (Kenilworth, NJ: Rodopi, 2003), Arne Vetlesen, *Evil an Human Agency: Understanding Collective Evildoing* (Cambridge: Cambridge University Press, 2005); Aaron Beck, *Prisoners of Hate: The Cognitive Basis of Anger, Hostility, and Violence* (New York: Harper, 2000); James Waller, *Becoming Evil: How Ordinary People Commit Genocide and Mass Killing* (New York: Oxford University Press, 1997).

Phenomena like this are understandable when it comes to cute robots like Boxie or entities like the Replika chatbot, where assigning intrinsic value is easier because we can have conversation and experience responsiveness. But what about death machines?

Death machines, of all the robots today, ought to be highly considered as objects of moral action, worthy of intrinsic and extrinsic value. These entities behave in inhumane ways as objects of violence and without a thought or care for the humans they destroy. They inhibit our most fundamental human emotion and impairment toward sin: aggression. If an entity is made to destroy a soul, an image bearer, it ought to be considered on the highest moral level that we can ascribe to it. I know this sounds absurd but hang with me here. Death machines are slaves to the minds of their makers, their freedom will most likely always be constrained. No one wants to create Skynet. Systems like Aegis and the Tomahawk missile, advanced as they are, are still simple, but the minds that made them are not. Death machines and human operators form a symbiotic relationship to accomplish their mission, this is driven by defense policy in the West, and most likely will remain a staple. DoD Directive 3000.09 (May 8, 2017), while vague, does state that human operators have freedom of the will over the machine or system.[203] It is also the standard, in the West, that these systems comply with Geneva Conventions and Rules of Engagement (ROE). Human-oriented practices of warfare are standard, as evidenced in the four standing Geneva Conventions, and International Humanitarian Law (IHL). Even though animals have served the US military over the years, and thousands of robots are deployed to ground, air, and sea, currently there is little to no consideration of their use or abuse.

Conclusion

Going back to the religious vision of DARPA, the reason why dogs, dolphins, horses, and now robots are considered as "military equipment" under Federal law (cf. Act of 1949 FPASA) is to ensure that the desired ends of the government are achieved. Even though emotional bonds form between handlers and both have risked their "lives" for one

203 Department of Defense, "Directive 3000.09," www.esd.whs.mil/portals/54/documents/dd/issuances/dodd/300009p.pdf, (accessed Nov 16, 2022).

another, they are still considered a tool.[204] Soldiers anthropomorphize or zoomorphize robots in the battlefield not because of the design of the entity but because of the human capacity for empathy and mercy. I do not believe this is going away, both because of the attachments to war and the symbiotic nature of human-machine warfare. Nor do I think organizations like the International Committee for Robot Arms Control (ICRAC) is going to bind and dismantle the creation and use of death machines, unfortunately. The ICRAC is fated to fail because it simply demands the status quo: human-centric control and consideration. The human operator is the main problem, for the Lethal Autonomous Weapon Systems (LAWS) did not create itself. Another pathway forward is to limit human control by granting limited rights and moral consideration to the object or as our friend David Gunkel states, "to consider the faceplate of the robot."[205]

The public has no power to restrain the desires of DARPA. However, we can push for the "rights" (or "rites") of things.[206] The vision of robotic futurism in the eyes of DARPA is that violent technology is the liberator of all humans at the yoke of the slave.[207] Another problem in this logic is that it utilizes violent technology to liberate one set of humans at the expense of another, making them into a Golem. Perhaps seeing the "I" in the machine is not a problem in the human psychological code, but a design by our creator. Not only that, but there is a desire for humans to connect with *things*. As Martin Buber reflects on this *Erfhrung* (experience), "man has a great desire to come into personal relation with things and to imprint this relation on them."[208] As we will

---

[204] Sarah Curse, "Military Working Dogs: Classification and Treatment in the U.S. Armed Forces," *Animal Law* 21, no. 2 (2015): 249–84; Peter W. Singer, *Wired for War: The Robotics Revolution and Conflict in the Twenty First Century* (NY: Penguin Books, 2009); Julie Carpenter, *Culture and Human Robot-Interaction in Militarized Spaces: A War Story* (NY: Ashgate, 2015).

[205] Gunkel does not argue for rights for (killer) robots per se, but that we challenge seeing them as only instruments. See "The Rights of (Killer) Robots" in *Moral Responsibility in 21st Century Warfare: Just War Theory and the Ethical Challenges of Autonomous Weapon Systems* (SUNY, 2020).

[206] Vilém Flusser, *The Shape of Things: A Philosophy of Design*, trans. Martin Pawley (London: Reaktion, 1999).

[207] Cf. Richard Coudenhove-Kalergi, *Eine Idee erobert Europa: Meine Lebenserinnerungen* [An Idea Conquers Europe: My Memoirs] (Vienna: Kurt Desch, 1958), 39.

[208] Martin Buber, *Schriften über das dialogische Prinzip* [Writings on the Dialogical Principle] Martin Buber Werkausgabe 4, ed. Paul Mendes-Flohr (Gütersloh: Güthersloher Verlagshaus, 2019).

see clearly in the next chapter, dehumanization through this technology is already a present reality, the moral injury is severe. The way forward is regulation not prohibition, to humanize the death machine, because we already see ourselves within it. It enacts our visions of colonization and political violence. It is both an extension of our humanness and our fallenness. To extend mercy to the *it* is to show mercy to ourselves. Otherwise, we are doomed to the political visions of violence and destruction that already entangle our addiction to tech and war.

When it comes to violence and its implements, I share the pessimism of Jacques Ellul. Propaganda, much like the logic of DARPA, is subject to what Ellul calls *technique*,[209] which ultimately ends in the development of such tech as death machines. It compels the public to believe that violence is a necessary force of protection and security, and that violent technologies are the solution in our turbulent post-9/11 world. Certainly, these threats are real and not just for those of us in the US, lest we forget about our fellow humans under siege currently in Ukraine. The reality of violence and aggression does not negate our ethical due care to balance security issues, privacy, and civil rights. For these violent technologies not only impact us within the context of war, but also within that of policing. Ellul and Arendt have noted quite well that police are the ultimate *technique*, for instance, the police during the Holocaust did not violate any laws under the jurisdiction of Himmler. [210]

I end this chapter with a calling to radical resistance to the vision and romance of DARPA. It is costly, but worthy if we desire to keep the peace. We have failed to show mercy to our human partners in the wake of political threats and unending violence, thus, the cycle continues. I am not suggesting that granting moral consideration to violent technologies magically solves the human condition of aggression. Yet, we know empirically we see ourselves inside the machine, even if there is "nothing" inside. Perhaps, if the machine can be an implement of political violence, it can also be an implement of peace, mercy, and love. Perhaps it should have the ability to turn the other *cheek* when its human operator would not. The *right* to refuse a launch or strike could be the

[209] Jacques Ellul, *The Technological Society*, 394.

[210] Jacques Ellul, *Violence: Reflections from A Christian Perspective*, trans. Cecelia Gaul Kings (Eugene, OR: Wipf & Stock, 2007), 86.

first step toward more ethically violent technologies. In that case we need to humanize and not demonize it.[211]

[211] Asher D. Biemann, "Humanizing the It: Martin Buber on Technology and the Ethics of Things," *Religions* 13, no. 137 (2022): 1–25.

CHAPTER 4

The Gaming of Death and Violence

[content warning: death, horror, gore, and dehumanizing language]

Introduction

While violent tech begins with enframing and ends with death machines, there is an intermediate state that experiments with the psychology of its user. Violence is entertaining. It is not just about the gore or shock of seeing simulated or enacted death without the fear or threat of harm. With the modern state of practical effects within film and the extreme realness that modern graphic processing units can emulate the physics of dismemberment and death, it is no wonder why the West is so entertained by virtual death. CEO of Activision, the makers of *Call of Duty* (CoD), recorded in 2021 that lifetime sales of the series reached 400 million. The online platform, *Battlenet*, averages about 100 million users monthly.[212] These numbers make it the most popular gaming series to date, and there is a reason for that. There is much more at play to the interworking of violent video games and their impact on the user, although it is most likely not what you have been told by mass media.

212 "Tisha Aploinario, "How Many People Play Call of Duty? (User & Growth Stats 2022)," *Fiction Horizon*, October 16, 2022, https://fictionhorizon.com/how-many-people-play-call-of-duty-user-growth-stats/, (accessed Nov 16, 2022).

Understanding the philosophy of video games, specifically violent ones, helps us further understand violent technology and its acceptance within Western frameworks. Let me just say out of the gate that not all virtual violence is the same. Much like the gradient presented in chapter 1, there is a great deal of variance in the violent video games on the market today. Context also matters a great deal when it comes to the actual psychological impact of the game. For example, take two very graphic (although very different) current games on the Steam market today. *Resident Evil 7: Biohazard,* released in 2017 (hereafter RE), is by far the most graphic and disturbing in the series of the franchise. The psychological horror experience during the gameplay is intense. From the mechanics to the eerie environments, the player is emersed within a world of relentless combat and limited resources. The developers of RE also designed the game from a first-person perspective, which is unlike other RE games. One of the features that makes this experience so terrifying is the constant looming threat of enemies the protagonist cannot destroy, at least until the boss battles later in the story. I will withhold spoilers in order to not ruin a very unique gaming experience. However, while the game is very violent, it is usually asking the user to engage in self-defense, granted at times extremely gory.[213] So, this is one tier of violent games, and I would put it on the shelf with many others that are acclaimed like *Outlast*, *Dead Space*, *Bioshock*, *Silent Hill*, and *The Evil Within*. Primarily, the user is operating (via the protagonist with the story arc) out of self-defense and survival.

There is another tier of violent video games that is more niche. Games like *Kindergarten* (Fig. 3), which lack the vivid detail of AAA games like RE but are disturbing and violent in a different way. For example, the description reads, "Welcome to Kindergarten! Have fun trying to learn and share… and not die. That is important, too. In a school that's just not quite right, it's important to be friendly…and take out those that aren't." A little vague, no? The object of the game is to figure out why the teacher and janitor are killing other children in the game. It channels the dark humor of T.V. shows like *South Park* and *Rick and Morty*. I place this game in a different tier because the violence is both unrealistic (i.e., pixelated) and obviously satirical.

[213] For those who are curious, see, "Resident Evil 7 Gore Compilation–Resident Evil 7 Goriest Moments–Resident Evil 7 Gory Death," *YouTube*, March 29, 2017, https://www.youtube.com/watch?v=PFD-Ym7_kQ0, (accessed Nov 16, 2022).

As disturbing as the concept of *Kindergarten* is, it is protected under the first amendment right of freedom of speech. Even still, this is not the most disturbing game I have either researched or experienced. Why? Unlike the typical violent video games that users experience, whether they be in the realm of senseless and satirical violence or in self-defense, there is another tier that seeks to prepare the user to respond to and rationalize violent behaviors toward other humans. Here I speak not only from research but from personal experience.

Fig. 3. Screenshot from Kindergarten. Photograph taken by author.

Dave Grossman, in his work *On Killing*, claims that there is a historical problem of getting infantry men to fire at the enemy. If this is true, it is a problem for organizations that are sending soldiers to do exactly that: kill the enemy. Grossman's book argues that there are psychological barriers to killing another human, but there are also means of overcoming these barriers. According to Grossman, only fifteen percent of soldiers fired their rifles at the enemy in World War II. This is staggering considering the amount of vetting the US War Department (now DoD) did before allowing men to enter service, rejecting over one million people. The public may be aware that during WWII 405,399 men were killed in action and 670,846 were wounded. But the public may not know that 1,253,000 service members suffered stress related injuries.[214]

Unfortunately, Grossman's research is questionable.[215] He bases much of his research on the work of S. L. A. Marshall, citing such work frequently without any formal citation in the endnotes. Marshall's work on the reluctance of the soldier to pull the trigger has also been seriously doubted in its historiography. Nevertheless, Grossman's primary thesis that one's physical distance from the target equals either a greater or lesser intensity to the psychological aversion to killing is further called into question by modern research on remote killing. Grossman, along with many other researchers mentioned above, have also put forth a faulty thesis that the gaming of violence equals desensitization to violence in real life. The gamification of death and its actualization in death machines play a critical role in the war-machine, but it has little to do, at least empirically speaking, with lessening the aversion to killing or even the desensitization of killing. Let us now turn to the psychology and philosophy of remote killing and all of its interlaced gaming. Where else should we start but with the philosophy of its design.

Design begins with a philosophical ideal. Philosopher of media, Vilém Flusser argued that design is about two things: identity & etymology. Flusser was quite pessimistic about the possibility of ethical design of technology. Why? The first reason relates to etymology. Flusser pointed out in *The Shape of Things: A Philosophy of Design*, the German *macht* and *mögen* derive from the Greek *mechnos* which means "to deceive" or "to trap."[216] The ontology of the machine is deception. What is not clear in Flusser's logic is whether the deception is banal or manipulative. I assume he would argue that it is a manipulative deception, since his own use of tech was very minimalistic; he did not write with a word processor, for example. Another key etymology that Flusser draws out is that the German for 'artificial' (*gekünstelt*) stems from the English word "cunning."[217] Thus, the philosophy of design for such things as AI,

---

214 See B. H. Chermol, "Wounds Without Scars: Treatment of Battle Fatigue in the U.S. Armed Forces in the Second World War," *Military Affairs: The Journal of Military History, Including Theory and Technology* 49 (1985): 9–12.

215 Dave Grossman, *On Killing: The Psychological Cost of Learning to Kill in War and Society*, rev. ed. (New York: Back Bay Books, 2009), 5–38.

216 Flusser, *The Shape*, 17.

217 Flusser, *The Shape*, 18.

robotics, and death machines, is designed to exploit the laws of nature and their creators. This manipulative deception would be impossible[218] for Flusser to label as ethically *good.* Another key aspect of design, according to Flusser, focused on the form and matter (i.e., identity of the object). He does critique war tech directly for this reason. For Flusser, who was writing under the growing tension of WWII, witnessed first-hand the political ambitions that were attached to the philosophy of design. For example, the submarine (e.g., *The Submarine*) was meant to be an object of world domination and ultimately utopia. As we noted earlier, once humans discovered the unity of matter and energy, the political ideologies in power could not keep themselves from trying to design implements of war and death. The submarine was designed to escape a flood of biblical proportion (cf. Genesis 6), but it failed. The philosophy of this design was not well received by the public. In essence, it would be like saying that world peace and harmony can be accomplished by drones with hellfire missiles. There is just something amiss about arguing that death machines equal peace. What does this have to do with the gaming of death? I believe it matters greatly, because the narratives the public believes about international conflicts, technologies, and their place in warfare shape both the design and the values we attach to them.

The Values Laden with the Gaming of Death

While I can appreciate Flusser's pessimism, it is not something I wholeheartedly agree with when it comes to the philosophy of design. Within violent tech, especially in the realm of gaming, there are redemptive elements that should not be overlooked in their design or identity. The argument that violent games lead to violent ends has been debunked. In fact, in many cases it leads to lower aggression; even being used as a treatment for PTSD.[219] The public misunderstanding about the design philosophy behind violent video games is only further eclipsed by the public misunderstanding about the military life and how it uses the gaming of death to prepare soldiers for the stress of combat. Remember we are talking about violent games that are not simply meant as entertainment,

---

218 See Patrick M. Markey and Christopher J. Ferguson, *Moral Combat: Why the War on Violent Video Games is Wrong* (Dallas, TX: BenBella, 2017).

219 Consider the research of Dr. Michelle Carras and the non-profit Stack Up. See https://www.stackup.org, (accessed Feb 25, 2023).

where games like CoD offer fast-paced combat at unrealistic engagement ratios, no ROE, and very narrow narratives of war. We will unpack all of these elements throughout this chapter, but before delving in, consider an alternative view of design philosophy that is not pessimistic.

Any good Christian theological philosophy must be filled with hope, and yes this even comes into play when dealing with death machines and gaming. Design is underlined by faith. Faith in what? Well, that depends on the designer(s). Ethan J. Brue, Derek C. Schuurman, and Steven H. Vanderleest, all engineers, put forth a modern vision for ethical design in their work *A Christian Field Guide to Technology for Engineers and Designers*.[220] According to these authors, through "design norms" a system can be modeled on a particular idea or requirement when it is validated and verified.[221] What they mean by this is the design is checked and balanced by asking two questions: (1) "did we do the right thing," and (2) "did we do the thing right?" Of course, this is based on a Christian assumption about what is right, moral, or good. What that means is that for the technology to be labeled as "good" it must be designed in a way that honors God. This is the normative principle or rule of measure for Christian engineers. In a secular framing of this norm, we could say that ethical design is one that honors human and environmental flourishing. Some modalities[222] (borrowing from Dooyewerd) that could be incorporated into this rubric might include the following:

Social	Does this design promote healthy social relationships, interactions, or roles?
Formative	Does this design have a positive impact on culture, history, and technology?
Economic	Is this design frugal?
Aesthetic	Does this design promote harmony and enjoyment?
Juridical	Does this design promote rights, responsibility, and due care?
Ethical	Is this design one of love and generosity?

Table 1. Dooyewerd's Modal Aspects of Reality

220 Ethan J. Brue, Derek C. Schuurman, and Steven H. Vanderleest, *A Christian Field Guide to Technology for Engineers and Designers* (Downers Grove, IL: IVP Academic, 2022), 66–92.
221 Brue, et al., *A Christian Field Guide*, 69.
222 Adapted from Dooyeweerd's Modal aspects of reality.

As we further examine violent tech in this chapter and the next, we must ask whether death machines meet these design norms or not. For now, the question of gaming design will suffice. In particular, I would like to consider First-Person Shooter (FPS) games used by the military and the public to communicate the narratives of warfare in the modern world. How might we use this tech to communicate the brutal, bloody, and boring aspects of war experiences? Not all modern FPSs are the same, and many lack depth and range in conveying experience. But it does not have to be that way and there are many redemptive components to these games that can help soldiers, law enforcement, and veterans alike.

The Gaming of Pain

Violence can be experienced on two levels simultaneously: physical and psychological.[223] The military, across the branches, seeks to build both mental and physical resilience through stress-induced training. The dimensions of stressors include duration, unpredictability, uncontrollability, intensity, and novelty. While these dimensions can all be found, in some respects, in violent games like *Call of Duty*, *Mortal Kombat*, *Hitman*, and so forth, there is a unique telos to the military-designed games: real-world application. This is an important distinction because members of the armed forces undergo training in order to learn adaptive skills that will translate to better decision making under the stress of real-life combat situations. The question remains, how effective are these simulations or games upon the mind and body of the soldier and is this violent tech morally good and acceptable?

This raises the question of whether virtual environments can affect physical states. On a philosophical level, whether you are a functionalist or not, one may not know the physical identity of pain, but one can know that being in pain alters physical behavior. In fact, the telos of pain is designed to do that very thing: behavior modification. For functionalists, like David Lewis, pain states are identical to physical states. Lewis developed this thesis in his 1966 article "An Argument for Identity Theory." In sum, Lewis claims that mental states are consistent in their

223 No, I do not believe that philosophically *pain* or *stress* has a rigid destination within the brain or that it is physically identical as a state. Also, within the doctrine of Army training and clinical research there is not a rigid identity either.

function within a system and not simply by their internal features. The causal role is what determines the type of mental state. Pain (or C-fibre firing) is caused by injury to a part of the body. The physical state is identified by the experience, and they are one and the same, according to Lewis.[224] Here we see the axiological assumption of materialism; that is, pain is rigid in its identity, and it occupies a role that is identified as 'causing pain.' Alik Pelman argues that the only way to challenge or verify this view of identity theory is by adding some minimal requirements. consider Pelman's proposition:

> *R*: "Pain" should designate, with respect to every possible world, all and only states felt as pain[225]

Pelman formulates the theory above to help access Lewis' view as well as the counter argument made by Saul Kripke who claims pain is nonrigid.[226] Kripke challenges Lewis' claim of identity by stating that if two rigid terms fail as an identical referent in a potential world, then it is logical that they fail to meet *R* in all possible worlds. Meaning that if it is possible for C-fibre firing to be distinct from pain then they are distinct.[227] While this is a battle of semantics, it is important. Both Lewis and Kripke formulate positions that meet requirement *R*. How do we explain the identity issue regarding the rigidity of pain? Much like the problem of other minds, (i.e., phenomenal experience), there is no way to solve the question of rigidity: if it feels like pain then it is pain.

I believe this question is important for the present study for the following reasons: First, the gaming of death (via machines) causes pain to the user and the human targets engaged. Close physical distance in killing (per Grossman's view) is not a requirement for the pain of violence to be real and psychologically hurtful. For example, research on drone operating teams bears witness to the fact that virtual killing, if you may, can lead to Post Traumatic Stress Disorder (PTSD).

---

224 David K. Lewis, "An Argument for the Identity Theory," *The Journal of Philosophy* 63, no. 1 (1966): 17–25.

225 Alik Pelman, "Metaphysics of Pain; Semantics of 'Pain,'" *Raito* 28, no. 3 (2015): 306.

226 Saul Kripke, *Naming and Necessity* (Cambridge, MA: Harvard University Press, 1980), 148–153.

227 Let the reader understand that this is no small controversy in the study of pain and semantics. Cf. Scott Soames, *Beyond Rigidity: The Unfinished Semantic Agenda of Naming and Necessity* (New York: Oxford University Press, 2002); and Joseph LaPorte, *Rigid Designation and Theoretical Identities* (Oxford: Oxford University Press, 2013).

Second, the telos of war is the infliction of moral injury alongside bodily injury. As a non-reductionist, I believe that while the semantics of pain are important, there is also wisdom in the relation of pain to the body-mind rather than in parts. Thus, we see the importance of the instrument and the operator–there is a link between them. The human operator becomes the causal force in the implementation of lateral pain and moral injury. Virtual environments become a necessary structure for the implementation and implements of pain and injury. But where does this process start? Where else but media, especially games?

Many are tempted to say, like Grossman, that the conditioning processes of the modern military take *normal* men and women and teach them to kill. If this is true, then it is possible to radically transform the human psyche within a matter of 9-15 weeks of basic military indoctrination. I do not believe that is the case at all. Rather, from experience, I know that what the indoctrination process does is draw out aggression that lies within. Media, like games and film, entice the user through a particular narrative to justify the moral injury of war. Elaine Scarry notes in her work *The Body in Pain* that there is a means to make this pain invisible. Two methods used in warfare to make pain invisible are *renaming* and *omission*, according to Scarry.[228] Enemies are called 'targets,' geographical locations of massive death are renamed to codewords, and the event of killing is labeled as neutralization. This process begins long before the soldier becomes an extension of their weapon system or vice versa; it begins at home with politicians and playthings.

Removing the political veils of *renaming* and *omission*, Carl von Clausewitz, in his tome *On War*, writes,

> It [war] is therefore for many reasons possible that the object of a combat is not the destruction of the enemy's force, that is, of the force immediately opposed to us, but that this only appears as a means. But in all such cases it is no longer a question of complete destruction, for the combat is here nothing else but a measure of strength–has in itself no value

[228] Elaine Scarry, *The Body in Pain: The Unmaking of the World* (New York: Oxford University Press, 1985), 69.

except only that of the present result, that is, of its decision.[229]

Regardless of the names we call the *enemy,* or which politician justifies the invasion and occupation of another country, the reality of warfare is two human operators seeking to enforce their will upon one another. This does not require the soldier to understand what that *will* is, but that they will be obedient to follow others unto death. It all begins with the belief that service to one's country is worth pain, moral injury, and potentially death. This is not a hard narrative to sell once the gaming of death and war have the cosmetic veneer painted by developers and Hollywood alike. Technical implements are crucial components to ensuring that the process of moral injury continue in the public imagination. Even before the allure of pixelated screens, painters like Pieter Bruegel the Elder (ca. 1525–1569), and Hieronymus Boscha (1450–1516) used art media to show the connection between tools and pain. Notice the connection between implements and pain in Pieter's *Der Triumph des Todes* ("The Triumph of Death"). The implements (i.e., swords, wagons, boats, sickles, axes, etc.) show the very good design of these technologies to enable death on larger scales. The goodness of these technologies is directly proportionate to the amount of death they allow through the extension of their human (or in these case animated skeletons).

Returning to the question of rigidity above, I do not know if mental states are identical to physical states, but the effects seem to be lateral upon both the body and mind. Human anthropology and archeology alike are witnesses to a creature that is in a constant flux of pain management and desire to avoid pain all together. Long before the utopic allure of the death machine, there existed a vision of a future where death and pain would exist no longer. The prevailing vision to enact this future always comes back to *force*, *violence*, and the damage and destruction of the *other*. Thus, the conditioning to move from a civilian who is not legally permitted to kill (except in justified self-defense) to a soldier does not require re-coding their psychological barriers, but instead grounding and justifying the use of aggression already present within their biology. If this sounds like Nazi eugenics, please bear with me. The study of

[229] Carl von Clauewitz, *On War*, ed. J.J. (Col.) Graham (Berlin: The Floating Press, 2010), 91.

neurocriminology suggests that there are some cultures, and individuals that are more prone to violence than others. From a military doctrinal standpoint, is to find those individuals and enlist them to take part in the machine.

Fig. 4. Pieter Bruegle's "The Triumph of Death" (c. 1562), Museo del Prado

Again, this is more related to economics. When it is proactive rather than reactive, violence literally pays off. Okay, but maybe those are fringe examples, what about everyday *domesticated* people in the West? Evolutionary theory does not do much to explain violence and criminal behavior; however, neuroscience has provided some evidence that aggression (proactive and reactive) is genetically predetermined. If this is true it gives new light to the theology of *original sin*, where we literally inherit the sin of our forefather Cain.[230] Through the study of identical twins, which are more genetically similar than fraternal, it was observed

[230] Technically Adam is the first human, but he is not begotten so much as he is made from the hand of God in Genesis 1. Cain, however, is the first *modern* man and therefore I have labeled him our *forefather*. Also, Adam in the Hebraic, and rabbinic literature is never mentioned in pejorative terms, unlike his counterpart Eve. It is not until the Epistles of Paul that theological readers have a negative view of Adam (cf. Romans).

that antisocial behavior is 40 to 50 percent explained by genetics.[231] Violence, on some level, is inheritable. But there is much more. Research conducted by Han Brunner revealed that a defective gene–Monoamine Oxidase A (MAOA) correlates to violent behavior, attention problems, alcoholism, drug abuse, and impulsivity. This also describes most of the combat arms units within the military branches of the Army and Marine Corp. [232] As Terrie Moffitt and Avshalom Caspi discovered, poor living and emotionally negligent environments also play a significant role in lower levels of the MAOA enzyme. Maltreatment is also associated with antisocial and violent behavior.[233] What does this have to do with the gaming of death and death machines?

A machine needs an operator. Where should the Military look for such operators? The primary target of recruitment is in lower income areas. Take kids that grew up in poverty, no options for college, or the troublemakers and they are prime candidates for enlistment and indoctrination. The genius of this process, which is informed by research, is that the media has done most of the leg work in providing the motivation for joining: to serve one's country. The target recruit would have lower levels of MAOA. Of course, there are no blood tests or PET scans to verify this inference, but there is clear heredity in the recruitment pool. Most likely recruits stem from military families. So not only is there a genetic potential for aggression, but there is a narrative driven by the prior service members that further solidifies the choice to serve.[234] No one, since the drafts of WWII and Vietnam, was forced to join the military; it is a voluntary service. Indeed, but the process is not apolitical either.

The grooming process starts with games and film media, which depict unrealistic and unbalanced perspectives of what war actually is in its nature. Films like *Saving Private Ryan*, *Lone Survivor*, *American Sniper*, and television series like *Terminal List*, *Jack Ryan*, and *The Selection* glorify both

---

[231] Raine, *Anatomy*, 41.

[232] H. G. Brunner, M. Nelen, X. O. Breakfield, H. B. A. van Oost, "Abnormal Behavior Associated with a Point Mutation in the Structural Gene for Monoamine Oxidase A," *Science* 262 (1993): 578–80.

[233] A. Caspi et. al., "Role of Genotype in the Cycle of Violence in Maltreated Children," *Science* 297 (2002): 851–54.

[234] Dave Philips and Tim Arango, "Who Signs Up to Fight? Makeup of U.S. Recruits Shows Glaring Disparity," *New York Times*, Jan 10, 2020, www.nytimes.com/2020/01/10/us/military-enlistment.html, (accessed Nov 16, 2022).

the warrior ethos and war-making. Handsome actors fight bravely to protect or work alongside a beautiful *femme fatale* and their country against a domestic or foreign threat. What is not often depicted is the impact of war on all societies involved and the mind, body, and soul of the war fighter. What eighteen-year-old is considering such questions? However, even the most elite (i.e., Tier 1: SEAL or Delta Force) soldiers come to grips with the reality of warfare. For a brutal and hilarious memoir of what warfighting is like, I suggest the interviews on the YouTube channel *The Shawn Ryan Show*. Ryan, a former Navy SEAL and CIA contractor knows intimately what happens on the battlefield and the process of the game upon the soldier. In a recent interview with Rob O'Neil, the SEAL that killed Bin Laden, there is great philosophical reflection by both Ryan and O'Neil about the reality of killing other humans for political reasons.

> O'Neil: People don't realize the humanity of it. These are dudes in their houses with their families. Okay, he is going for a gun, he's got a gun and he is aiming at me, but is he protecting his family? Is he a carpenter? Does he need an AK because he lives in f****** Ramadi? And again, I think it's because…the more senior that I got, the one thing I disagreed with was patrol to contact and strike to develop. It's like, look man, you are going to go to a house and find someone with a gun and kill them because you went to it? I remember I went into the wrong house, I was in a swamp, and there was this white carpet, and the only people in the house were a woman and her daughter. I looked down at the carpet, and I remember thinking, I know why they hate us. Imagine someone doing that here?

This transcript from a three-hour discussion shows the dual nature that many elite soldiers must occupy. On one side they are human: fathers, sons, and members of a brotherhood; and on the other side able to flip a switch and engage in brutal acts of conflict. SEALs, Rangers, Green Berets, and Delta Force are made up of highly qualified, vetted, and intelligent members. We cannot reduce them to being more aggressive than other members of society. Neither can we say that they were brainwashed by the media into trying out for the most elite outfits in the West. The harsh truth is that the West does not exist without these men (for better or worse). They did not create the machine or its ideology, but they are the *good* design of the war-machine, and they are glad to stand for what they believe in: freedom.

Most of the members of these Tier 1 groups have solidified this career path early in their teenage life. It is a career path they have sought out early on. For example, those wishing to attend BUDS (Basic Underwater Demolition School) must first join the Navy and go through Basic Training. There are no guarantees. It is the same with the Army Qualification course for Special Forces. At the end of these grueling schools, even if the attendee makes it through the *process,* they may not be picked by the cadre to advance to teams. What makes someone put themselves through this and knowing that best-case scenario, they will be deployed to high conflict areas for both combat and humanitarian missions? It is the warrior ethos that enables them to play this game, knowing that eventually they might lose. Make no mistake, the military reminds its recruits from reception forward that they signed a contract and made an oath to the Constitution of the United States, and all health and wealth of the individual are on the table when the call comes.

Shannon French, in her excellent book *The Code of the Warrior*, explores this phenomenon throughout history.[235] Romans, Vikings, Monks, and Samurais frame the ancient ethos of a warrior code that death is better than dishonor. To further clarify, it is not merely the honor of the soldier, but the honor of the symbols the soldier wears, the tabs on his or her shoulders, and the patches that represent their time in combat. All these symbols represent the bonds that take place in battle and boredom. Media representation is what pushes certain individuals to play the game of death.

There is a rich tapestry that makes up the psychology of a warrior. The warrior ethos cannot be reduced to deficiencies in enzymes (i.e., MAOA) and poor living environments. What drives people to risk their life and the life of others? It might surprise the reader to know that it is a rag-bag mixture of vanity, honor, heritage, and pragmatics. People join the military for money, to experience war, because their father was in X or Y war, or to pursue experience in a particular field. Reductionist approaches to warrior ethos ignore this reality and the long heritage of codes that they belong to. Army training doctrine understands this and

[235] Shannon E. French, *The Code of the Warrior: Exploring Warrior Values Past and Present* (New York: Roman & Littlefield, 2017).

from in-processing (the time you get off the bus) until you receive the DD-214 at the end of service, the goal is to bind selfish individuals into a unit of selfless warriors (somewhere between hero and coward). One effective way to accomplish the transformation from civilian to warrior is through gaming. Leadership courses, virtual interactions, and extreme death by PowerPoint (where the soul dies) encode these values of teamwork and bear witness to the consequences of those that fail to do so.

I bring this up because there is a temptation to dehumanize the warrior and demonize their implements. Neither of which relieves a grasp of the morality or psychology of modern warfare. The games (past and present) of warfare are not simply about killing or even desensitizing the soldier to killing, but rather giving them and their leaders an assessment of their behavior under stress. Playing and planning through war-gaming is not comparable to playing FPSs war games. To make such a comparison is to show how uninformed one is about the nature of training through violent technologies and the nature of war itself. If modern game-makers and gamers want an *authentic* experience, there is a lot of work to be done. This does not go away in the face of the death machine or remote killing. In fact, gaming provides a new layer to the culture of killing.

Controlling the 'Killbox'

Unlike the visions of postconditioning and indoctrination put forth by Grossman, the games military play with violent technology is more about affective than reflective violence.[236] Research into the nature of drone and remote killing reveals that spatial distance does not equal psychological distance. In other words, the mediation of violent implements does not make the act less traumatic or more ethical. The intensity of remote killing is equally traumatic in perception.[237] Training simulations like *Killbox* and *Virtual Battle Space* (VBS) unfold the entanglement between gaming and remote warfare. But the mediation of

---

[236] John P. Aizaid and Rosa Brooks, "Recommendations and Report of the Task Force on US Drone Policy," *Stimson*, April 2015, www.stimson.org/wp-content/files/file-attachments/recommendations_and_report_of_the_task_force_on_us_drone_policy_second_edition.pdf, (accessed Nov 16, 2022).

[237] Hugh Gusterson, *Drone: Remote Control Warfare* (Cambridge, Mass.: MIT Press, 2016); Wayne Chappelle et al., "An Analysis of Post-traumatic Stress Symptoms in Unites States Air Force Drone Operators," *Journal of Anxiety Disorders* 28, no.5 (2014): 480–87.

war-gaming is nothing new considering the emergence of video games (c. 1960s).[238] What is *new* is the biopolitical myth or acceptance of violent technologies to mediate global conflicts. It is *new* in the sense that for the first time in history, nations have the political power to extend the gamification of death from war-tables to the battlefield. The 9/11 conflict juxtaposed to the success of many military-based games normalized militarism in the United States. A virtual war on the big screen and computer monitor kept the urgency of avoiding technological surprise ever present.[239] Good military politics and training are about streamlining war and battle-ready soldiers. Again, this begins long before the trainee enters Basic Combat Training (i.e., Bootcamp). Training environments and indoctrination processes are not about brainwashing or fundamentally changing the individual from civilian into cold-blooded killer. Rather it is to place the human in affective atmospheres and unveil whether the human is violent or not. If the individual is not naturally aggressive, training helps them through affective environments (and muscle memory) learn to follow simple commands, yes, even killing another human. The Army does not download a *kill code* through U.S. Army and Doctrine Command (TRADOC) into the trainee, humans are not machines. Capacities for violence and aggression are natural. Even though most people believe they are *good*, that has nothing to do with whether they would kill. Given the right atmosphere and condition we can all be violent. Neuroscience of violent behavior concedes with this thesis: given the right social environment and genetic predispositions, it is not difficult to game violence or death. Violent technologies are not the causal agent behind reflexive or proactive aggression, but they do provide affective textures to soften the reality of the *killbox*. Enough of political and social commentary, let us move on to analyze the games themselves.

I have mentioned only two games thus far in the chapter: VBS and *Killbox*. Yet there are other games like *America's Army: Proving Ground* that are critical to the war-building complex and its embeddedness within US culture. While there has been much critique about the nature of games

[238] See Nina B. Huntemann and Matthew Thomas Payne, *Joystick Soldiers: The Politics of Play in Military Video Games* (New York: Routledge, 2010).

[239] Robertson Allen, *America's Digital Army: Games at Work and War* (Lincoln: University of Nebraska Press, 2017); David Neiborg, "Training Recruits and Conditioning Youth: The Soft Power of Military Games," in *Joystick Soldiers*, 53–66.

like *America's Army*,[240] there are also positive aspects to such games because they provide affective environments for ethical tactics. Granted, every game is different and provides a variation in play, educational delivery, sophistication, and immersion. Each one provides a different experience. *America's Army* is what one might expect from a recruitment tool, it allows the gamer to play an 11B (infantryman) within the context of a Long-Range Combined Arms-Recon (LRCA-R) unit. The game requires that the user qualify with their weapon system (e.g., 240B) and learn basic tactics and techniques that those going through LRCA-R training would know. Typical of FPSs of this flavor are fast paced "true-to-life" missions, weapon systems, and aesthetics. This is obviously a subjective rendering of "true-to-life" because anyone who has ever served in combat theaters knows there is a difference: boredom, waiting, preparation, more waiting, fatigue, pain, trauma, and brief moments of excitement. Few military games reflect these realities of war.

Prior to a deployment to Iraq, the Army used a training software known as VBS. The soldier and his squad would enter a massive floor to ceiling 360-degree gaming simulation to recreate a brief mission in theater. I had the joy of going through this training in 2010, and it is nothing like the experience of playing an FPS. Andrew Barron, through the investments of Bohemia Interactive Simulations, found a way to apply affective environments for soldiers training for deployments. My experience, like many other soldiers, was a mixture of entertainment and frustration. That has nothing to do with the game, its designers, or the poor civilian operators who were assigned to a bunch of frustrated soldiers being forced to play war. In fact, the gameplay was made to be realistic, requiring the soldiers to occupy a mock Humvee and AR-15s that had been modified to respond to the soldier's engagement with targets on screen. However, as one can imagine, when these systems are used in high volume and in constant need of calibration, things do not always work like they should. For example, the brakes on our Humvee during the mission went out and we all died when the driver decided to run off a local bridge. Up-armored Humvees are not known for their

[240] James Der Derian, *Virtuous War: Mapping the Military-Industrial-Media-Entertainment Network* (New York: Routledge, 2009); Robertson Allen, "Software and Soldier Life Cycles of Recruitment, Training, and Rehabilitation in the Post-9/11 Era," in *The War of My Generation: Youth Culture and the War on Terror,* ed. David Kieran (New Brunswick, N.J.: Rutgers University Press, 2017).

crash safety ratings. When the equipment did work, many of us had a hard time not shooting the local wildlife, running over wildlife, or disrespecting the ROE. Before you judge, remember, we had been training for months with no time off and minimal outlets for frustration. Needless to say, command and civilians alike get frustrated when lower enlisted do not comply.

I bring this up because it shows the reality of war for *real* and remote warfare. The humor is dark, the boredom is never-ending, and when there is a chance to relieve frustration, we take it. Environments that game warfare ought to reflect this reality. Operators of the MQ-9 Reaper (i.e., drone) experience this as well. It is a test of attention, affect, and aggression. Drone pilots sit in a small trailer (8x30) for twelve-hour shifts. In front of them are four screens that monitor fuel, maps, weapons, and all of which are on a continuous feed; there is no "pause" button. Games like *Killbox*, developed by Joseph DeLappe, Malath Abbas, Albert Wert Elwin, and Tom Demajo, allow the player to assume two different roles: a drone pilot targeting the Waziri village in Pakistan or a Waziri villager. As can be imagined, the role of the pilot is one of immense power. Under this role you experience command to command prompts that ultimately end in a missile strike: a clinical and disconnected experience for the pilot. The role of the villager is quite different.

Unlike the drone pilot, the villager has unlimited freedom, but no power. Their telos in the game is to be the victim. To experience anxiety, the shaking of the ground, and to choke on the billowing smoke is the fate of this village. *Killbox* depicts the killing of drone strikes quite well, but the scope of the experience is limited. The experience is limited for at least two reasons. First, both the operator and villager experience are fully immersive. Between modern LIDAR and IR sensors used in the MQ-9 Reaper, the details in both killing and observation are three-dimensional and graphic. Granted, *Killbox* is not about ultra-realistic textures, rather it is about education. The drone operator is limited in their *free will*. Commands are dictated to the kill team and the user must *feel* the tension embedded within these scenarios. The incorporation of the villager perspective is helpful in this regard, but it does not go far enough. The primary purpose of the psychology of the gaming of death is to test the user in affective climates. Second, the emulation gives the user an immersive experience because the strike is the long-awaited climax to an otherwise painfully boring existence: surveillance. Prior to

2001, the drone was about eliminating technological surprise and providing low risk/low cost to espionage. In modern warfare, the drone is an extension of the human operator as both a means of surveillance and target engagement.

Modern FPSs are fatalistic. Linear narratives force the player to bend to the will of the non-playable characters (NPCs). Many of these games, particularly *Modern Warfare* and *Black Ops*, operate without a concern for ROE, focusing rather on old tropes (i.e., war is hell, we did what we had to do, and so on).[241] A major myth within these games is that killing is not a serious and responsible business. Whether through virtual environments, Reapers, or from the cockpit of an A-10, the gaming of death and war is not to be taken lightly. The conception the public holds of the modern battlefield matters greatly.[242] I speak specifically from and to a Western perspective, but the by-product of the war-machine in the West affects the entire world. Through new media like video games, Netflix, and even Twitter, the modern citizen has intimacy with war unlike ever before, and apparently it sells quite well. Virtual intimacy does not equal physical intimacy, which shows in the myths evidenced in media deceptions of warfare. Physical distance is not the only reality divorced from modern depictions of war in media. As any researcher of the warrior ethos will attest, there is much more to the warrior than a desire to pull a trigger.

My concern about the virtual misrepresentation of warfare in modern gaming is echoed in the research of Andrew Barron, director of design of Bohemia Interactive Simulation. In 2018, Barron presented a presentation titled, "Depiction of War in Games: Can You Do Better?" which addressed these very concerns. As a veteran of the Afghanistan war, Barron has first-hand experience with war and the depiction of it in games. In the conference presentation, Barron quotes a former Marine's realization that modern warfare is not like *Modern Warfare*, stating "this

---

[241] Call of Duty: Modern Warfare, 2019, *Infinity Ward*, Activision; Call of Duty: Cold War, 2021, *Infinity Ward*, Activision.

[242] Christopher Coker, *Waging War Without Warriors: The Changing Culture of Military Conflict* (Blouder: Lynne Rienner, 2002); Neil Renic, *Asymmetric Killing: Risk Avoidance, Just War, and the Warrior Ethos* (Oxford: Oxford University Press, 2020); Juliane Ravel, and Madalena Rosa, "IHL in Action: Seven Patterns of Respect," *Humanitarian Law and Policy*, December 2, 2021, https://blogs.icrc.org/law-and-policy/2020/11/19/ihl-in-action-seven-patterns-respect/, (accessed Nov 16, 2022).

isn't what I expected. I thought I would be doing hero s*** like in Call of Duty."[243] Barron goes on to tell the developers present in the crowd how important the depiction of war in games is to public expectation. Barron is right. For example, a brief survey of the *Call of Duty* gaming experience presents an unbalanced view of warfare.

In *Modern Warfare*, the "war is hell" motif is evidenced early in the episode of the Piccadilly Circus. The motif is displayed in an exchange between characters Price and Kyle:

> Price: You saved lives today, Sergeant
> Kyle: This shouldn't have happened in the first place. They sent us in half-ass so everyone could keep pretending we're not in a war.
> Price: Yea, lie of the f****** century, that is.
> Kyle: Then why have we got our hands tied. Let's take the bloody gloves off and fight … we don't stand a chance in hell with these rules of engagement, sir.

Similar logic is applied in *Black Ops: Cold War* released in 2020. One of the leading characters, Adler, continually says, "We've got a job to do" to the user's player Bell. This "job" constantly involves torture, no ROE, or any regard for humanitarian approaches to law within warfare. Hopefully, this myth will be less prevalent in future FPS games.[244] Critics of *Modern Warfare* have called out these shortcomings and are hoping to see more FPSs that promote "realism" or a deeper integration of the status and rule of law in warfighting.

The "war is hell" motif (*inter arma silent leges*); that rules do not and should not apply to battlefield applications, is a favorite of the CoD franchise. In games aimed toward the public, the limits and rules of war fall short. This is not the reality of warfare. However, former US secretary of State, Donald Rumsfeld gave a statement in 2001 that reveals that "war is hell" is a competing narrative in defense logic. Rumsfeld, when questioned about collateral damage responded,

---

243 Andrew Barron, "Depiction of War in Games: Can You Do Better?" *Global Gaming Conference,* March 2018, https://mail.google.com/mail/u/1/#inbox/FFNDWMkjVFHsRxfVrkXRjGQZtTKvBXvh?projector=1, (accessed Nov 16, 2022).

244 Neil C. Renic and Sebastian Kaempf, "Modern Lawfare: Exploring the Relationship between Military First-Person Shooter Video Games and the "War is Hell" Myth," *Global Studies Quarterly* 2 (2022): 1–11.

> There has never been a conflict where people have not been killed, and this is the case here. There are ordinates flying around from three different sources. It's flying around from us from the air down, it's flying around from the Al Qaeda and the Taliban up those lands somewhere and kills somebody when it hits, and there's opposition forces and Al Qaeda forces that are engaged in shooting at each other. Now, in a war, that happens. There is nothing you can do about it. We lost 5,000 people in this country, plus. And we need to stop people, terrorists, from doing that. They're the ones that started this conflict. They're the ones that imposed great damage on the Afghan people. They're the ones–the Al Qaeda--that have invaded Afghanistan with a foreign presence that ought not to be there at all. Do I think that that is a worthy cause? You bet I do. And will we stick to it? You bet.[245]

Rumsfeld's sentiment is a common narrative within warrior ethos: many sins of war can be pardoned in the name of defense. Under the Trump administration, many war criminals (i.e., former Blackwater contractors) were pardoned.[246] Even though these actions undermine the law of war, it is also evidence of the seriousness of moral and legal responsibility within military culture, even if the president does not respect it. As I have mentioned earlier, there is no such thing as clean killing, but neither does the "war is hell" myth fully capture the game the military is playing. Seeing this logic firsthand, it is understandable why media, especially video games, would incorporate these myths into their narrative. That does not make creators less responsible or guiltless under the veil of freedom of speech. Rather, the public should consider the critiques of the legal, gaming, and military community about what warfare is and is not like.[247]

[245] Donald Rumsfeld, "Text: Defense Secretary Donald Rumsfeld," *The Washington Post*, October 29, 2001, www.washingtonpost.com/wp-srv/nation/specials/attacked/transcripts/rumsfeldtext_102901.html, (Accessed August 14, 2022).

[246] Adela Suliman, "Trump's Blackwater Pardons Erase the Line Between Slaughter and Justified Wartime Violence," *NBC News*, Dec 23, 2020, www.nbcnews.com/news/world/trump-s-pardon-blackwater-contractors-convicted-massacre-greeted-derision-grief-n1252224, (accessed Nov 16, 2022).

[247] Cf. Renic and Kaempf, "Modern Lawfare."

Asymmetrical warfare, that of drones and other violent implements of war, gives rise to a symbiotic relationship between operator and instrument. Meaning, there is a great deal of power in remote killing and warfare, but that does not automatically mean it is disenchanted.[248] That is to say, the use of police and military drones does not necessarily equal a "rule by Nobody" (Arendt) or the Hobbesian Leviathan.[249] The same was essentially said of the Submarine during the wake of WWII. It is easy to forget, at least in public memory, that machines require mechanics (i.e., maintenance), and a master. The minerals and metals that make up these implements are forged from a fallen (or cursed) ground.

No matter the fatalism of one's political theory, its success or failure always requires and assumes that humans will fall in line to serve as both operators and maintainers. Remote killing is accomplished in the same way that mechanized killing via tanks, aircraft, or aquatic vessels has always been done by integrating human and machine. Disenabling the machine is *killing* the team, so to speak. For example, Joseph Chapa, a pilot and philosopher, notes in *Is Remote Killing Moral* that "a pilot becomes an ace with five air-to-air kills against the enemy…The fighter ace's five or more air-to-air kills refer not to enemy pilots but to enemy aircraft."[250] With the Predator and its successor the Reaper, there is a little more to the political game at hand: it becomes a political theology. Much to my surprise, there is very little theological reflection on the metaphysics and theology of drones. I assume this has much to do with the acceptance of JWT, but nevertheless, the lacuna is telling. Thus, the remaining section of this chapter will seek to fill in the gap.

The drone, like any other implement, is not so much a disembodied weapon as it is a weapon to mediate distance between a target and an operator. The drone strike still impacts the mind and body of the operator 7,000 miles away. The drone is maintained by several teams, one in country and one out of country. It takes both to make the implement operational. Theologically speaking, remote killing should not be the only aspect addressed. I think the language, training, and political systems of the *game* of remote killing are just as troubling. Let us start with the

[248] See Christopher Trigg's critique in "Drones, Hobbes, and Liberal Enchantment," *Political Theology* 19, no. 7 (2018): 533–71.

[249] Shaw, *Predator Empire*, 9.

[250] Joseph O. Chapa, *Is Remote Killing Moral? Weighing Issues of Life and Death from 7,000 Miles* (New York: Public Affairs, 2022), 46.

language. Words matter. God created the world with words of blessings and goodness. If we are going to suggest redemptive qualities to the drone program, perhaps it is best to examine the language.

Much like the process of violence, the usage of drones begins with a particular epistemology. One that obscures the nature of the human receiving violence and death at the hands of the operator and technology. Consider the following terms taken directly from the DoD Dictionary:

> **adversary** –A party acknowledged as potentially hostile to a friendly party and against which the use of force may be envisaged. (JP 3-0)
> **casualty** –Any person who is lost to the organization by having been declared dead, duty status – whereabouts unknown, missing, ill, or injured. (JP 4-02)
> **close support** –The action of the supporting force against targets or objectives that are sufficiently near the supported force as to require detailed integration or coordination of the supporting action. See also direct support; general support; mutual support; support. (JP 3-31)
> **collateral damage** –Unintentional or incidental injury or damage to persons or objects that would not be lawful military targets in the circumstances ruling at the time. (JP 3-60)
> **execution planning** –The Adaptive Planning and Execution system translation of an approved course of action into an executable plan of action through the preparation of a complete operation plan or operation order. Also called EP. See also Joint Operation Planning and Execution System. (JP 5-0)
> **target** –1. An entity or object that performs a function for the adversary considered for possible engagement or other action. 2. In intelligence usage, a country, area, installation, agency, or person against which intelligence operations are directed. 3. An area designated and numbered for future firing. 4. In gunfire support usage, an impact burst that hits the target. See also objective area. (JP 3-60)

These terms are but a small survey of how the game of war is played: dehumanizing the enemy. Yes, sometimes this means a literal dehumanization, as in seeing the other human as *less than* or *sub-human*. Even in the TRADOC document above, we see a softening of the reality of killing another human, by calling them a "target." Other times the veil is more obscure as with the term "execution planning." When decoded, this means that those in power calculated the losses of human life, budget, and the concerns of loved ones in light of the mission. The incorporation of this language into military usage reveals that the defense department has built an epistemology that enables the use of death machines in a veiled way. Calling a human, a target, does not make the process easier. PTSD is still on the table, but it makes the dehumanization of others a daily part of conversation. The drone or death machine, as an implement, does not enable or encourage this, but it does become a byproduct. It is not merely the death machine, but also the algorithms that justify to higher command the usage of these machines.

John Emery, Assistant Professor of International Security in the Department and Area Studies at the University of Oklahoma, brings this to light in his critique of ethical due care.[251] Emery reminds the public that the military has used algorithms to predict collateral damage in potential drone strikes: the Collateral Damage Estimation Tool (CDET), or *bugsplat*. Emery's concern, and I share it, is that by reducing ethical due care to computation lends itself to unjust usage of the violent tech we employ in warzones. Using this tool is a way to simply "check the box" of ethical due care in order to force the will of one set of humans upon another. Allowing this reality is based upon the linguistic and epistemic foundations built from TRADOC forward. I believe this current linguistic approach to military doctrine is ethically, morally, and theological problematic.

In classical Christian Theism, the doctrine of the *perichoresis* communicates the metaphysical and ontological reality that humans are members of a body–connected to one another and to God. The Eucharist is, as Louis-Marie Chauvet mentioned in *Symbol and Sacrament*, a sign of the "presence of absence." Borrowing the phenomenology of Heidegger (i.e., *I-body*), Chauvet, argues that it is through the body we are situated, linguistically, and holistically. Chauvet tells us that we are interwoven:

---

[251] John R. Emery, "Probabilities Towards Death: Bugsplat, Algorithmic Assassinations, and Ethical Due Care," *Critical Military Studies* 8, no.2 (2022): 179-197, DOI: 10.1080/23337486.2020.1809251.

> The I-body exists only as woven, inhabited, spoken by this triple body of culture, tradition, and nature. This is what is implied by the concept of corporality: one's physical body certainly, but as the place where the triple body–social, ancestral, cosmic–which makes up the subject is symbolically joined, in an original manner for each one of us according to the different forms of our desires.[252]

The Eucharist is what brings us into the "language-game of faith."[253] Karl Rahner agrees with the logic of Chauvet here that the Eucharist is but one media that Christ uses to communicate and mediate intimacy with the church. What about the axioms in Scholastic theology (cf. Lateran Council of 1215) about transubstantiation? I appreciate Chauvet here because the mediation is not about the media (i.e., bread and wine) but the symbols found with the fissure of the bread and drinking of the wine–being broken. This game we play in accordance with the command of the Lord (Matthew 26:17–29; Mark 14:12–25; Luke 22:7–38; and I Corinthians 11:23–25). The language and practice are about bringing the absence of God into this present reality before the presence of the community.

I am not sure that TRADOC is aware of the insights of Lacan and Heidegger about the process of mediation, but they do employ the antithesis of this philosophy. The drone and death machine, as we see above, is built on an epistemic framing that is about anti-presence. Even before we get to the critique of the implement, there is a deep theological challenge to the telos of the death machine via the *perichoresis* or the Latin *circumincession* which translates "a proceeding around."[254] In its nature, the death machine is about mediating the presence of the human team operating it. It stands as a symbol of death and destruction in the presence of image bearers. This is not a moral commentary on the service

---

[252] Louis-Marie Chauvet, *Symbol and Sacrament: A Sacramental Reinterpretation of Christian Existence*, trans. Patrick Madigan, SJ, and Madeleine Beaumont (Collegeville, MN: Liturgical Press, 1995), 150.

[253] Chauvet, *Symbol and Sacrament*, 378.

[254] A. Deneffe, SJ, "Perichoresis, circumincessio, circuminsessio. Eine terminologische Untersuchung," *ZKT* 47 (1923): 497–532; [G.] L. Prestige, 'Περιχωρέω and περιχώρησις in the fathers," *JTS* 29 (1928): 242–52; G. W. H. Lampe (ed.), *A Patristic Greek Lexicon* (1968): 1077, s.v. περιχωρέω.

members that employ and utilize these technologies, but rather about the political theology that makes these implements of judgment. Based on this logic alone, the Church is to be an anti-drone community. But there is still more to say on this system that games death.

The dehumanization of other humans is only one component of this game. As a former service member, I can speak from experience that the games we play both in training and in war can mentally and physically leave scar tissue. The weapon is a symbol of pain, constantly reminding the service member they have committed to a bigger machine. Trust in the American Military has drastically lowered since 2019 when recruiting exceeded the Army's expectations. Following the pandemic of 2020 (and continuing in 2023), retention rates are at a low that the Army has not seen since Vietnam. Why? Despite efforts to use social media, film, and cringe worthy commercial campaigns about how the Army values individuals, the public is seeing the truth: global security has a domestic cost. Remember, as stated earlier, Yarden Katz's monumental work, *Artificial Whiteness*, states there are epistemic forgeries that the public must believe in order to keep the war machine functioning. The public is waking up to the fact that the biomedical and security promises of a global war on terrorism are nothing short of idolatry. The promise that death machines make war *clean*, more accurately, and ultimately more just are potential realities for those 7,000 miles from the battlefield, but even then, psychological analysis of remote pilot PTSD would argue otherwise.

Currently, the US Military is way under its goal of retaining a 1.3 million manned armed force. While this is encouraging in the sense that the public is seeing the true cost of maintaining a global military presence, the downside is that more and more the military is looking into AI and robots as a substitute. That is not to say that the armed forces are seeking to remove humans from the equation, that is not and never will be the case. Rather what is happening is the human-machine team is further being solidified and the human as the remote killer is becoming more normative. This means greater pain and traumatic exposure to a smaller number of already over-worked service members. What can the public do about this problem?

Prohibition is not the answer, as tempting as it is to proffer that this is the way forward. Civilians play violent games for therapeutic reasons. While that might seem disturbing to those that find violence in video games revolting (and certainly there are games that warrant such), there is data that communicates positive aspects of violence in games. The release of violent AAA games like *Grand Theft Auto* correlates to few homicides.[255] There is a catharsis to violent games. Virtual pain and death have physical manifestations upon the user or gamer. The Virtual Environments Laboratory at University College London replicated the study of Stanley Milgram's famous prison experiment, and the results of the study were very similar. Volunteers encouraged to shock prisoners, even in the wake of their screams and cries for mercy, exhibited signs of physical stress. Virtual environments are perfect for providing affective textures for the study of human behavior. The cries from public critiques have been that virtual violence desensitizes the user, but the opposite can also be true: it cuts both ways. What if we develop games that endeavor, through virtual mediation, to develop virtue and morality?

Moral Injury and Redemption

Echoing the concerns of Andrew Barron and Joseph Chapa, virtual environments and remote warfare require as much ethical due care as traditional warfare. Regardless of where we fall on the use of violent tech like Reapers, the Christian community has an obligation to care for the service member. This means ensuring that the training environments help them analyze how they behave under stress, caring for their mind and body during and post-war, and helping them reintegrate into a society that may be foreign or apathetic toward their experiences. After all, Jesus said in the Gospel of John, "no greater love a man has than this, that he would lay down his life for his friend." It is a respect for human dignity that theological communities must advocate for, that regardless of which side of the implement they are under, they matter. When I speak of redemption in the realm of gaming and death machines, I do not have hope that the framing of war and attachments modernity has to warfare will have redemption. I remain a pessimist on those matters. I realize there is a great deal of literature about the possibility of nonviolence and the future of war, but I will reserve comments until the next chapter for those matters.

[255] P. M. Markey, C. N & J. E. French, "Violent Video Games and Real-World Violence: Rhetoric Versus Data," *Psychology of Popular Media Culture* 4, no.4 (2014): 1-19.

In the book of Isaiah there is a beautiful passage about the redemption of violent tech:

> And He will judge many nations and render decisions for many peoples; and they will **shape** their swords into plowshares and their spears into pruning hooks. Nation will not **take up** sword against nation, and never again will they **exercise** war.[256]

This is a literal rending of the Hebrew apparatus from Isaiah 2:4, but we might also put it into a modern framing by translating it in a paraphrase:

> And they will shape their AR-15s into gardening tools, and their drones into therapy robots. Nation will not use violent technologies against one another again, and never again will they war game.

The prophet Isaiah is referring to the eschaton when the implements of war will no longer have purpose. The extent of atonement is not merely a future reality, for healing and repair are possible in the here and now. Through virtual reality, social robots, and immersive gaming environments, could we not use implements that were once means of violence for peace and recovery? I certainly think that this is a real possibility, but the nature of the game as well as its impact must be understood by developers and caregivers.

My primary concern about the gaming of death through technologies is the moral injury they cause. This is not to be conflated with PTSD and the debates around remote pilots. Moral injury is something different and extends to service members regardless of whether they have used drones or not.

There are many definitions of moral injury, but the most to date and comprehensive definition comes from J. D. Jinkerson,

> Phenomenologically, *moral injury* represents a particular trauma syndrome including psychological, existential, behavioral, and interpersonal issues that emerge following perceived violations of deeply held moral beliefs by oneself

---

[256] Author's translation. Words in bold translated to capture the sense of the Hebrew root. Isaiah 2:4.

> or trusted individuals (i.e., morally injurious experiences). These experiences cause significant moral dissonance, which if unresolved, leads to the development of its core symptoms.[257]

Discussing violent tech in light of moral injury is helpful because it revolves around doubt and not mere combat experience or physical injury. Studying moral injury also helps frame the study to come in our next chapter in the critique of JWT and its moral code. Brad E. Kelle, in his work *The Bible and Moral Injury,*[258] does a wonderful job of analyzing episodes within Scripture that describe the process of moral injury and biblical warfare. Kelle's work is unique within the realm of this literature because his focus is not solely on the individual but instead on the communal aspects of moral injury. For example, in chapter 4, Kelle surveys the postwar rituals in the Old Testament that helped the soldiers deal with the shame, grief, guilt, and social disconnection they experienced. Kelle notes that these Old Testament rites and rituals include (but are not limited to) the following:

- The purification of people and objects (including captives): Num 31:13-24.
- Erecting memorials and monuments: Ex 17:14–16; Num 31:48–54; Josh 6:24; 1 Sam 5:1–8; Dan 1:1–2; 5:2–3.
- Communal Lament: 2 Sam 1:19–27; Pss 44; 60; 74; 79; 80; 89.

Kelle uses moral injury as a heuristic to read these texts alongside current psychological research, offering insights for modern veterans and those traumatized by war and violence. It is great to see such careful and considerate research on such a murky topic and by a theologian, nonetheless. What is missing from the analysis, and I do not fault Kelle for this, is a proper framing of the warrior ethos prior to the conditioning of the indoctrination process and the experiences of warfare. As mentioned in chapter 1, the process of violence does not begin or end with violent experiences, but rather with desire. Kelle brings in the

---

[257] J. D. Jinkerson, "Defining and Assessing Moral Injury: A Syndrome Perspective," *Traumatology* 22 (2016): 126. Emphasis original.

[258] Brad E. Kelle, *The Bible and Moral Injury: Reading Scripture Alongside War's Unseen Wounds* (Nashville: Abingdon Press, 2020).

research of Camilo Bica and Pete Kilner,[259] both of which follow the definition of *moral injury* stated above. What I think is missing in both accounts, and the literature overall, is the realty that most recruits (18-24) lack a formally developed moral framework.[260] Placing the blame of moral injury heavily on the environment of training or war is a mistake. Basic training does not *undo* the nature of the individual; it does condition and modify certain behaviors, but that is not the same as an "eroding moral being" as Bica asserts. I am not saying that soldiers are never forced or commanded to do things that might contradict their *felt* beliefs, but it is too strong to say that it is this environment alone that leads to injury, desensitization, and replaces their basic beliefs and values. This also goes against the research of those like Shannon French in tracing the long history of warrior ethos that seems to be ingrained in every culture. The evidence from history is that some men and women can use their violent nature for the benefit of the civil and military industrial complex. Some of the most *violent* men in the Western hemisphere (i.e., numerous confirmed kills) are very kind and gentle friends and fathers. As evidenced in countless hours of interviews with Tier 1 warriors, they volunteered for elite units out of a sense of honor and many times brotherhood. There is also a particular personality trait that manifests among these warriors: extreme selflessness. As much respect as I have for the work of Belle and others around moral injury, perhaps they misunderstand the game that is being played. For many of these elite soldiers, the moral injury began when the events of 9/11 happened. As Shawn "Machine" Lavery has voiced in a candid interview, "once I was able to process what I was seeing, I was angry … I didn't grow up in a military family, so I didn't have this sense of patriotism burned into me as a child, but by the time I was in college, 18 years old at this point, I

[259] Camillo Bica, *Beyond PTSD: The Moral Casualties of War*, War Legacy Series, vol. II (Commack, NY: Gnosis, 2016); Pete Kilner, "Moral Injury: Personal, Organizational, and Divine (Perceived) Betrayals," *Thoughts of a Solider-Ethicist* (blog), Nov. 1, 2016, http://soldier-ethicist.blogspot.com/2018/01/causes-of-moral-injury-perceived.html, (accessed Nov 16, 2022).

[260] "Characteristics of Active Component Accessions Age," *Prhome Defense*, https://prhome.defense.gov/portals/52/Documents/POPREP/poprep99/html/chapter2/c2_age.htm#:~:text=Approximately%2C%2087%20percent%20of%20new,age%2021%20(11%20percent), (accessed Nov 16, 2022).

was very proud to be an American … I felt like the planes were flying into my body physically was the level of frustration I had."[261]

Seeing the virtual image of the planes collide with the Twin Towers on 9/11 was a massive and collective event of moral injury to future service members. One that sent many young men and women toward years of training, deployments, and service, for which many gladly gave their lives and limbs. I was in the eighth grade during that event and little did I know how it would change my life forever. Basic training did not undo my moral values or erode my moral being, it was a test of whether my desire to fight in the war would be validated. Post 9/11 there was no dearth of recruits seeking to join the military and benefit from its massive pool of resources. For myself and many others whom I served with, joining the military was about this quasi-religious devotion to the *polis*, some vague sense of obligation to the state which was further encouraged through massive marketing campaigns. But that does not mean there is not inner turmoil and wrestling with the frustration of military life. And it is here that I believe gaming could add some value through realistic affective environments.

When people find out that you served in the US Army, they often have these glib ideas about what that means. They might ask, "did you kill someone," or "what was basic like,"? But there is so much more to it than this reductionism. Yes, killing, violence, and death are a part of war, but so are humanitarian efforts, unconventional warfare, intelligence gathering, education, and finding belonging within a broad demographic makeup. How do we, as the public, ensure that these men and women integrate back into a civilian lifestyle? It is much harder than one might expect.

Moral Injury is just as prevalent among service members who transition to civilian life. People might say "thank you for your service," but that is about the extent of the value the public communicates. The betrayal, depression, and disillusionment that follow service separation has been called *living in limbo*, and rightly so. Qualitative research reveals that there is much ambiguity in this transition, which unfortunately can lead the former

---

261 "Nick "Machine" Lavery–Active-Duty Special Forces Amputee," *The Shawn Ryan Show* (*YouTube*), episode 28, June 20, 2022, 31:40–45, www.youtube.com/watch?time_continue=1905&v=k9JONFJp9dE&feature=emb_logo, (accessed Nov 16, 2022).

service member to despair or suicide.[262] The military is clearly aware of these issues and through clinical psychology has implemented two programs: "Impact of Killing" and "Adaptive Disclosure."[263] These courses reaffirm the thesis of Belle that communalization is important, if not necessary, for healing and repair. My question is whether gaming can add to this process of repair and redemption?

Redemption in FRAME

To accomplish this endeavor, we must start at the beginning with the media's depictions of warfare, primarily with the developers of these games. Communal repair is a biblical ideal, but it is incarnationally difficult for those who have been through trauma, especially veterans of war. The process of framing war begins with the developer, and the public must demand that they think about the ends of their gaming. If the goal of the FPSs is military realism to communicate the narrative of the warrior ethos, then there is much work to be done.

There is an acronym for everything in the DoD, perhaps there is poetic justice in considering an acronym for redemption–that is FRAME. The acronym FRAME stands for fidelity, responsibility, accountability, maturity, and efficacy.[264] It is a moral framing of the warriors'

---

262 Jan Grimell, "Revisiting Living in Limbo to Illustrate a Pastoral Psychological Understanding of Transition from Military to Civilian Life," *Pastoral Psychology* 68 (2019): 393–405; A. Adler, M Zamorski, T. W. Britt, "The Psychology of Transition: Adapting to Home After Deployment, in *Deployment Psychology: Evidence-Based Strategies to Promote Mental Health in the Military*, eds. A. B. Adler, P. B. Bliese, and C. A. Castro (Washington, DC: American Psychological Association, 2011), 153–74; D. Blackburn, "Out of Uniform: Psychological Issues Experienced and Coping Mechanisms Used by Veterans During the Military-Civilian Transition," *Journal of Military, Veteran, and Family Health* 3, no. 1 (2017): 1–8; M. Bragin, "Can Anyone Here Know Who I Am? Co-Constructing Meaningful Narrative with Combat Veterans," *Clinical Social Work Journal* 38, no. 3 (2010): 316–26; Alyssa J. Mansfield, Randall H. Bender, Laurel L. Hourani, and Gerald E. Larson, "Suicidal or Self-Harming Ideation in Military Personal Transitioning to Civilian Life," *Suicide & Life-Threatening Behavior* 41, no. 4 (2011): 392–405.

263 David Wood, *What Have We Done: The Moral Injury of Our Longest Wars* (New York: Little, Brown, and Company, 2016), 234–50; Brett T. Litz et al., *Adaptive Disclosure: A New Treatment for Military Trauma, Loss, and Moral Injury* (New York: The Guilford Press, 2016).

264 I have taken this from the work of Duane Larson and Jeff Zust in *Care for the Sorrowing Soul: Healing Moral Injuries from Military Service and Implications for the Rest of Us* (Eugene, OR: Cascade, 2017). Cf. Chapter 6: The Role of Moral Senses in Directing and Judging Military Character.

professional goals. Unlike other framings (i.e., Bica) where the warrior goes through a process of undoing their moral framework, FRAME allows the public to see that it is more of an interlining of their moral sense. The moral senses are the mechanisms that engage the muscles and mind of the individual. One's behavior can be conditioned and modified through external force, but the internal is much more complex. Unlike many FPS games that claim realism, there are no linear guides in real combat. One's inner moral voice and moral sense are all one has in discerning whether to obey or disobey an order. For example, Dick Couch, a former SEAL reflects on this tension when he notes his decision to disobey an order during his tour in Vietnam:

> I did not feel right about it. My men applauded the decision…I debate the order to this day. There was no question in my mind that the order was stupid and ill-conceived, one that would have put my men at needless risk. The work we did was dangerous enough without inviting disasters and getting caught at night in the mangrove swamp by a large VC [Viet Cong] force was just that. You either ran, fought, or died–sometimes all three. But I had sworn to obey the lawful order of "those appointed over me" and that day I refused . . . my men counted on me not to put them at risk unless it really counted. They had followed me into some hard situations that tour. I owed them a lot…I don't really have a sure answer for this one. I didn't then, and I still don't. Had there been a Junior Officers Training Course for SEAL officers in my day, as there is now, I might have made a better decision. Or a more informed one.[265]

Gaming needs to reflect this tension in the narratives they portray to the public, because it is a decision future soldiers will encounter. Responsibility and accountability are also extremely important concepts that are often neglected in military FPSs. Actions within war happen on a global scale. Therefore, responsibility and accountability are critical to the warrior ethos. Public trust and professionalism impact the injuries of

[265] Dick Couch and William Doyle, *Navy SEALs: Their Untold Story* (New York: HarperColins, 2014), 103–5.

combat. Developers have an ethical due care to represent this to the public through their narratives and mechanics.

Another component of the FRAME is maturity and efficacy. Environments of war and violence can distance moral agency from moral voice. This is why violence is mostly understood as a gradient. In a general sense, warriors must learn to decode what they are taught, whether right or wrong, and develop the capacity to teach and learn from past experiences of failure. In moments of stress, no matter the affectual environment, ethics can be thrown to the wind. Our moral voice might justify the use of force because we are afraid of harm, but that does not mean it will be justified in the eyes of authority. Neither does justification in the eyes of authority make it morally right or wrong. When it comes to *justified* use of force or violence it is always a case-by-case basis. The benefit of affectual environments is that games allow for post-event reflection and learning. The military already takes advantage of these environments, as we have seen earlier, but perhaps the civilian world could also implement such programs to account for a world of violence and depravity.

I refer to the importance of this training in the context of law enforcement, and the lack of funding that is required to educate our officers. Could not VR, gaming systems, and social robots be a cost-effective alternative to training in order to avoid the poor planning and execution of situations like the murder of George Floyd? Programs and training are not insurance policies but means to show the civilian and the warrior that without maturity and efficacy, physical and moral injury are the result. To tie in the biblical rationale about meekness here, Matthew 5:5 offers some ancient wisdom for a world of superpowers and technology.

Jesus, in the Sermon on the Mount, gave to the disciples a lesson on focus and control of their moral voice. He said, "Blessed are the meek, for they shall inherit the Earth." The word πραεῖς ("gentle, meek, humble") beckons the ethical usage of Aristotelian qualities about the use of force. In Aristotelian ethics, gentleness or meekness means a well-measured ability to control one's wrath.[266] However, for Jesus, there is an eschatological vision in mind. Yes, the LXX usually renders πραεῖς as

[266] Aristotle, *Nicomachean Ethics*, trans. Robert C. Bartlett and Susan D. Collins (Chicago, IL: University of Chicago Press, 2012), 4:11.

"gentle" or "meek," but the powerlessness is more about restraint for the sake of the future. Ps 36:11 in the LXX says that the "meek will inherit the land." What is meant here and in Matthew is not that the meek are powerless from an inability to use force or violence but that they willingly limit their power for the sake of the Kingdom of God (cf. Matt. 11:29; 21:5). It is about one's ability to restrain from the use of their power in the face of immediate benefit to the individual or even the benefit of the state. In light of the moral voice and agency of Jesus, perhaps what the Christian community needs to reconcile is our inability to limit our power.

The Western Christian vision of *freedom* is closely tied to a commitment to power–a self-seeking mission that disregards the physical and spiritual realities of the *other*. This is not just about warfare and the implements of war, but a commitment to making hell on earth at the cost of the soul. The incarnation of Jesus during the height of a military power like the Roman Empire bears witness to what he preached about submission and peacemaking in the face of power. Armed forces and their implements are not the ultimate reflection of the blessed life, in fact, the evidence on a global scale is that the cost of war is too high.[267] We will turn to consider the plea for anti-violence shortly, but for now and in conclusion, let us consider the final piece of redeeming the gamification of death.

Technology is eschatological; there can be no doubt at this point. The question remains though, where do we want to shine the brightest light of hope? Hope is undefeatable. That is why the object of one's hope matters for this life and the life to come. The promise of redemption, from a Christian theological perspective, is not about armed forces, implements, or political power. Nor is the promise of peace and justice reliant upon human logic or ideals. From media depictions to military campaigns, however, the promise before the public is that security and peace are the results of war-making. Nowhere do we see this logic more present than in the FPS that lures the player into a narrative of deception and death machines.

Gene Sharp in his book *Civilian-Based Defense: A Post-Military Weapons System* asked the reader to

---

[267] Cf. "The Costs of War Project by Watson Institute for International and Public Affairs," *Watson Institute of International and Public Affairs*, https://watson.brown.edu/costsofwar/, (accessed Nov 16, 2022).

> Imagine how successful war would be if there were no prior organized armies, no training of soldiers, no development and amassing of weapons and ammunition, no studies of military, no preparations of officers' corps, no arrangements for transportation and communication, and no emergency supplies of blood and provision for medical services.[268]

Sharp is alluding to a nightmare in the political vision of many conservative Western eyes, but one that is backed by research that states nonviolent campaigns outperform violent ones at a ratio of 2:1.[269] Not only does Sharp's point match the logic of the Sermon on the Mount but proffers the success of what Jesus called the 'Blessed' life. Sharp is referring to military campaigns and not all circumstances of violence, and I believe this is also true of the ethics that Jesus communicates as well. However, before the public can match moral agency to moral voice, there must be a voice communicating the truth of what Debos, Sharp, Stephan, and Chenoweth are communicating about the cost of war-making. Where is the media that communicates the possibility of Civilian Defense Systems, nonviolence, and so forth? One potential bridge into the mind of the public is gaming, both those in civilian and non-civilian sectors. If we believe that *freedom* and individual autonomy are part of the blessed life, then we must stop communicating a fatalistic narrative that the military industrial complex is the only place to set our hope in physical, mental, and spiritual redemption.

Conclusion

Video games can be powerful technologies to explore narratives. I wonder if, like the creation of the blues, games could provide affect environments for the reorientation of our past and present pain, suffering, and hopelessness. Globally one in three women have experienced sexual or physical violence.[270]Although a virtual

---

[268] Gene Sharp, *Civilian-Based Defense: A Post-Military Weapons System* (Princeton, NJ: Princeton University Press, 1990), 9.

[269] Maria J. Stephan and Erica Chenoweth "Why Civil Resistance Works: The Strategic Logic of Nonviolent Conflict," *International Security* 33, no. 1 (2008), 8.

[270] "Violence Against Women, Prevalence Estimates, 2018," *World Health Organization*, March 5, 2021, www.who.int/publications/i/item/WHO-SRH-21.6, (accessed Nov 16, 2022).

environment is by no means a path toward forgiveness and reconciliation, and we dare not put that on the victim-survivor. Rather, perhaps environments could be created for them to process their trauma and find an alternative avenue for the revenge they might seek. Games like *Depression Quest*, *Night in the Woods*, *Hellblade: Senua's Sacrifice*, *Gris*, and *Celeste*[271] are being used with therapeutic advantages. While I do not believe such games or virtual environments are appropriate replacements to therapy and reconciliation, they might be a valid supplement to aid the victim-survivor on their path to reconciliation.

[271] Aimee Daramus, "Video Games Can Be Used as a Therapy-Here's How," *Psychreg*, Jan. 20, 2020, www.psychreg.org/video-games-therapy/, (accessed Nov 16, 2022).

CHAPTER 5

Just War Theory and Violent Technologies

> "Some trust in chariots [drones], some in horses [tanks]...we trust in the LORD." Ps. 20:7
> "For punishment occurs only in the relation of a superior to those subject to him, and states do not stand in that relation to each other." Kant, *Metaphysics of Morals*, The Doctrine of Right, 57.

The use of violent technologies assumes that just war is possible. At last, we have come to the darker side of the gradient of violence–warfare. Violence may begin with desire and devotion, but the cycle usually ends in larger conflicts. To say conflicts end should be taken with a grain of salt, because it does not have a commencement, only further compels the cycle of violence to continue. Much like the narrative around national security and defense systems, Just War theory (hereafter JWT), either *jus ad bellum* or *jus in bello*, narrative needs critique in light of current political theories of violence and modern technologies. In JWT 'thinking' there is not a continual line of logic from the ancient Near East to DARPA. However, there are versions of the theory that the modern reader will recognize in the generic formulation of the doctrine. It is difficult, if not impossible, to examine the development of JWT without beginning with the Christian tradition. Our views of violence, and especially warfare, are usually situated in a historical and political context.

For the early Christians, post the crucifixion of Jesus Christ, pacifism was persuasive because the Kingdom of Heaven was imminent. The apocalypse of John, also known as Revelation, conveys this theology with impressive visual recapitulation of Hebraic thought. The forthcoming end of days and the final judgment posited to the Christian community to put hope in salvation and not through the means of the sword. Additionally, the book of Revelation is clear that at the end of days, the final conflict is between humanity and God.[272] At least this was the political hope of Christians during the reign of Domitian and Nero where scapegoating was a common fate of those that followed Jesus of Nazareth. But as history revealed, the avoidance of violence within larger political conflicts arose. This became even more complicated as the Roman Empire *converted* to Christianity under Constantine in 313 AD.

Those that argue that violence is necessary usually do so as a means of justifying it. Once we step into the cycle of violence (via warfare), we cannot escape its continuity, nor its reciprocity. "All who take the sword will perish by the sword," Jesus tells us in Matthew 26:52 (i.e., violence begets violence). But this is not where the literature of JWT often begins, it does so with the biopolitical narrative "kill or be killed," and thus we must rather focus on the *jus ad bello* or *jus in bellum*. So, does that mean JWT is passé?

This question is haunting for every man and woman that has experienced the trauma of warfare: separation, hostile environments, seeing the land die, witnessing the fragility of life, and the physical acts of violence upon others. I dare not claim to solve the complexities of JWT here, but there is one component that I wish to proffer for deeper consideration in the conflicts and wars to come in our age and beyond.

Reconsidering Just Peacekeeping

In the realm of institutional violence, such as warfare, it is the result of two conflicting desires and no formal authority to mediate the dispute.

---

[272] In the message to the church at Pergamum, John records the words of Jesus saying, "Therefore repent; or else I am coming to you quickly, and I will make war against them with the sword of My mouth" (Rev 2:16). As so often is the case, John echoes the Old Testament within his writing, and in the passage above the episode in Num. 22 is reverberated. Cf. G. K. Beale, *The Book of Revelation: A Commentary on the Greek Text*, New International Greek Testament Commentary. Grand Rapids, MI; Carlisle, Cumbria: W.B. Eerdmans; Paternoster Press, 1999), 250–51.

Traditional JWT positions itself to function as a judge (unofficially) between parties; and thus, according to Francisco di Vitoria is a way to return public order.[273] So, a *just war* is one of authority? Whence does this authority derive? Does it come from governments and generals? Where do we delimit what counts as a *war*? We have, in the wake of 9/11, an ongoing "war on terrorism" that occludes any formal start or end, which makes the *jus post bellum* another crisis altogether. An estimated 150 million to 1 billion people have been killed in wars throughout history,[274] who is the arbitrator of whether this is *justified*? These are questions that are outside the scope of this chapter; however, they are questions that deal with the morality and legality of what happens in framing justifiable violence and its impact upon society.[275] The present task here is to examine and critique violent implements.

Theologically speaking, JWT is permissible, hence the long-standing tradition, but that does not mean it is above critique or without blemishes. There is a great chasm between *jus ad bellum* and *jus in bello* and when discussion of implements comes into play the cavity is even greater. As we have seen in previous chapters, the framing of violence matters for domestic and foreign conflicts. Critics of JWT have pointed out that *just* does not always mean *good* or *right*; the details and languages we use matter. Michael Walzer, in *Arguing about War*, makes this criticism of the often-employed Machiavellian realism in the *just* war language that frequently appears in philosophy, politics, clergy, and military personnel. Walzer argues that at the beginning of JWT our focus is on the immediate issues and this is the wrong way to begin.[276] There is a similar framing in the discussion of violent tech, that is, should we ban LAWS or not. But

273 Francisco di Vitoria, *De iure belli relectio*, 3.9.60, in *Vitoria: Political Writings*, ed. Anthony Pagden and Jeremy Lawerence (Cambridge, MA: Cambridge University Press, 1991), 327.
274 Chris Hedges, "What Every Person Should Know About War," *The New York Times*, July 6, 2003, https://www.nytimes.com/2003/07/06/books/chapters/what-every-person-should-know-about-war.html, (accessed Nov 17, 2022).
275 Michael Gelven, *War and Existence: A Philosophical Inquiry* (Pennsylvania, PA: Pennsylvania State Press, 1994); J. T. Johnson and G. Weigel, *Just War and Gulf War*, Ethics and Public Policy Center, eds. J. T. Johnson and G. Weigel (Ethics & Public Policy Center, 1991); J. McMahan, "Proportionality and Necessity in Jus in Bello, in *The Oxford Handbook of Ethics of War*, ed. S. Lazar and H. Frowe (Oxford: Oxford University Press, 2016); Oliver O'Donovan, *Just War Revisited*, Current Issues in Theology (Cambridge: Cambridge University Press, 2003); James M. Dubik, *Just War Reconsidered: Strategy, Ethics, and Theory* (Lexington, KY: Kentucky University Press, 2018).
276 Michael Walzer, *Arguing about War* (New Haven, CT: Yale university Press, 2004), xi.

there is so much nuance that must go into the conversation before we can frame a technology *just*, *good*, *right*, or *unjust*. Therefore, John Langan, advocates for a wider spectrum of ethical consideration in JWT, beginning with *ante bellum* and *post bellum*.[277] Current theories that consider violence and war between the spectrums of pacifism and just war are in danger of tunnel vision. Glen Stassen makes a similar argument in his introduction to *Just Peacemaking: The New Paradigm for the Ethics of Peace and War*.[278] Addressing the root theological and philosophical concerns that lead to the cause of an armed conflict is critical for avoiding the moral vacuum of traditional JWT. A *Just Peace Keeping* perspective holds both the pacificist and military enthusiast accountable to a set of practices that delves into a holistic examination of conflicts. It is not so much an ethic as it is a correction to the temptation to swing the pendulum from one side to the other. Stassen recommends ten practices:

1. Support nonviolent direct action.
2. Take independent initiatives to reduce threat.
3. Use cooperative conflict resolution.
4. Acknowledge responsibility for conflict and injustice and seek repentance and forgiveness.
5. Advance democracy, human rights, and religious liberty.
6. Foster just and sustainable economic development.
7. Work with emerging cooperative forces in the international system.
8. Strengthen the United Nation and international efforts for cooperation and human rights.
9. Reduce offensive weapons and weapons trade.
10. Encourage grassroots peacemaking groups and other voluntary associations.[279]

These practices seek to make *just ante bellum* a permanent part of the ethical framework for what is considered a just and unjust war. Adding this paradigm into a just war framework also explores the complexities of international law. Steven Haines, academic lawyer and former naval officer, notes that this helps shape training doctrine for those preparing

[277] John Langan, "An Imperfectly Just War: Part of Uncertainty is Factual," *Commonweal* 118, no. 11 (1991): 361–65.

[278] Glen Stassen, ed., *Just Peacemaking: The New Paradigm for the Ethics of Peace and War*, rev. ed (Cleveland, OH: Pilgrim Press, 2008).

[279] Stassn, *Just Peacemaking*, 22.

for armed conflict. Haines writes, "the moral dimension of preparation for war might lead us to the identification of moral obligations which, in turn, could eventually lead us to the development of a set of formal rules."[280] From a theological perspective, the *telos* of war and conflict should aim for peace. This is the object of Augustine in the *tranquillitas ordinis*[281] and the ethical command to "love one's neighbor." Ethical due care is not a checklist and failure on either side of the conflict (i.e., *just ante bellum* or *jus post bellum*). Imperfect justice is the bookend to all conflict, but the key here is understanding the limitation of earthly justice. Just as with the critique of the reformed view of the fall and violence, which makes rationales for violence that may propagate such, the lesser of two evils (i.e., Niebuhr) view of just war can undercut its morality. It seems there is a tension in many theological circles, regarding JWT, that seeks a compromise between the nonviolent ideals of the New Testament ethic and the violent realism of modern anthropology. The prayer of the church should not be "let us act justly in war," but "Lord, help us to abolish the heart that longs for war."[282] Much like the DoD, modern scholarship on JWT needs to reconsider the moral language used in the discussion of violence and war.

Perhaps, as Mark J. Allman and Tobias L. Winright suggest, there is a spectrum of degrees when it comes to considering the justness of conflicts. Working off Langan's thesis about the imprecision of criteria they suggest examining the imperfectness of justice as degreed:

Degree	*ante Bellum*	*ad Bellum*	*in bello*	*post Bellum*
Perfectly Just War: Ideal and has never occurred.	X	X	X	X
Imperfectly Just War: Limited infractions	X		X	X

[280] Steven Haines, "Just Ante Bellum: Myth or Reality?" Unpublished Conference Paper, New York University Law School, November 9, 2010.

[281] Augustine, *City of God*, bk 19, ch. 12.

[282] This is not just a problem for Protestants. Cf. Darrell Cole, "Good Wars," *First Things* 116 (2001): 27; Stanley S. Harakas, "The Morality of War," in *Orthodox Synthesis: The Unity of Theological Thought*, ed. Joseph J. Allen (Crestwood, NY: SVS Press, 1981); *Wholeness of Faith and Life: Orthodox Christian Ethics: Part One–Patristics Ethics* (Brookline: Holy Cross Orthodox Press, 2004).

UnJust War: Violates or ignore criteria in all categories but *ad bellum*		X		
Perfectly UnJust War: Fails all categories				

Table 2. Degrees of Just War adapted from Allman and Winright

A perfectly just war has never occurred throughout human history, and most likely never will. Yet these categories help assess the moral theories we practice in relation to warfare. The ultimate *telos* of just war should be peace among nations, but perfect peace will only come in the eschaton. However, that may be, but we ought not to resort to a utilitarian rubric of what justifies our conflicts and how we use implements of violence in said conflicts. It is through the lens of just peacekeeping I recommend we examine violent technologies in the remainder of this chapter.

Immoral or Just Weapons?

The DoD created a Joint Artificial Intelligence Center in 2018 to study its adoption by the Military. This statement should be confusing to the reader because we know from the history of DARPA, that AI, computer science, and technological innovation of surveillance and remote warfare have been implemented by the military long before 2018. A more honest statement would be the DoD rebranded its take on AI in 2018 by proposing to use AI for US "security and prosperity."[283] As early as page 5, the document throws out the fear of reprisal upon the reader by stating that "Other nations, particularly China and Russia, are making significant investments in AI for military purposes."[284] How is this "new" threat any different than the emergence of nuclear warheads or biological warfare; both of which the US has led in using upon the enemy in their "just" wars? One major difference is that the violent technologies of the past assumed a human operator or decision-maker. As far as official policy,

[283] US Department of Defense, "Summary of the 2018 Department of Defense Artificial Intelligence Strategy: Harnessing AI to Advance Our Security and Prosperity," 2018, https://media.defense.gov/2019/Feb/12/2002088963/-1/-1/1/SUMMARY-OF-DOD-AI-STRATEGY.PDF, (accessed Nov 17, 2022).
[284] US Department of Defense, "AI Strategy."

humans remain the ultimate decision maker when it comes to systems that use AI. So, why the concern about autonomy?

Autonomy

Modern weapon systems have limited autonomy. Implements like landmines, heat seeking missiles, and even more advanced systems like the C-RAM (Counter Rocket Artillery and Mortar) or CIWS (Close-In Weapon System), have the autonomy to engage targets. Autonomy is a growing concern in the uses of AI in violent technology because of the nature of AI, Machine learning, and Big Data. The 2018 DoD statement is a reverberation of the third offset strategy (30S) and the concern about autonomy. In sum, the lines of effort (LOE) seek to bring to fruition the Cold-War vision of preventing technological surprise. In the logic of the DoD, per the 3OS, autonomy relates to the system to select and decide a course of action.[285] The strategy has never been to use AI in a general way; it has always been in *narrow-task* systems and human-machine collaboration. Narrow-task systems have some autonomy; as with the C-RAM, the weapon can identify and engage targets, but it relies on the human proxy for the final decision. The goal was human-level paced decision making and not necessarily flawless decisions. If machines were at least on par with the human proxy, the advantage for operational tempo was successful. The current plead by the DoD is human-machine teams or human-in-the-loop decision making, for good reasons. While LAWS are cause for concern in the public eye, there is a lot of misunderstanding. Just because a system is autonomous does not mean it is synonymous with AI. The AEGIS system used by the US and Japanese navy has the capacity for full autonomous engagement, but it is unlikely to operate in "human on the loop." Sea Hunter is a robotic ship,[286] developed by DARPA, that can autonomously hunt other

---

285 Bob Work, "The Third U.S. Offset Strategy and Its Implications for Partners and Allies," Jan 28. 2015, Willard Hotel, Washington, D. C., www.defense.gov/News/Speeches/Speech/Article/606641/the-third-us-offset-strategy-and-its-implications-for-partners-and-allies/, (accessed Nov 17, 2022).

286 "Sea Hunter: Inside the U.S. Navy's Autonomous Submarine Tracking Vessel," *Naval Technology*, May 3, 2018, https://www.naval-technology.com/analysis/sea-hunter-inside-us-navys-autonomous-submarine-tracking-vessel/, (accessed Nov 17, 2022); Joseph

submarines. Could it also be used to seek and destroy other objects or intelligence gathering? Of course. Kargu-2 is a drone that can use facial recognition software to engage targets autonomously, but that does not mean it will be used in that manner? Unlike video games and film media, there is a great deal of accountability for the service member working within these human-machine teams. Briefings, planning, ROE, Geneva conventions, and lots of red tape are in place by the DoD to ensure that if things go wrong there is a human in the loop to blame, punish, and correct.

The DoD is committed to AI (and likewise robotics) because it is an enabling technology. Autonomy is not consciousness and no one in DoD community is making such a claim. There is a saying in the Air Force, "data drives decisions" and thus we see here why AI and even LAWS are useful in armed conflict. It is not so much about taking the soldier's physical presence out of the fight, but about giving them better intel and tactical advantage. Deep learning, machine learning, and therefore AI is of great value to armed conflict because it frees the human operator of the rote activities of warfare. For example, Project Maven, a collaboration with DoD and Google developed automated pattern recognition algorithms to sort through video footage at speeds that far surpass human capabilities.[287] These systems and their software are seen as tools of intelligence and aids to those deployed in war zones. Anyone familiar with intelligence and reconnaissance within a military context understands the complexity and often rote nature of planning and implementing missions. The task is to observe, orient, decide and act, known in the military community as the OODA loop developed by John Boyd. When we automate and add AI systems into the loop it allows for more fluid, but not flawless, planning. Incorporating AI into the OODA

Trevithick, "Navy's Sea Hunter Drone Ship Has Sailed Autonomously to Hawaii and Back Amid Talk of New Roles," *The Drive*, February 4, 2019; www.thedrive.com/the-war-zone/26319/usns-sea-hunter-drone-ship-has-sailed-autonomously-to-hawaii-and-back-amid-talk-of-new-roles, (accessed Nov 17, 2022); Kris Osborn, "New Navy Ocean Attack Plan–Combined Air, Surface, and Undersea Drones," *Warrior Maven*, June 5, 2019, https://warriormaven.com/sea/new-navy-ocean-attack-plan-combined-air-surface-undersea-drones, (accessed Nov 17, 2022).

[287] Cheryl Pellerin, "Project Maven to Deploy Computer Algorithms to War Zone by Year's End, *"Department of Defense*, July 21, 2017, https://www.defense.gov/News/News-Stories/Article/Article/1254719/project-maven-to-deploy-computer-algorithms-to-war-zone-by-years-end/, (accessed Nov 17, 2022).

loop does not take away human agency. The AI software might excel at pattern recognition and coalescing large sums of data, but it does not substitute human critical thinking in the planning and implementing process, especially when it comes to adjusting the plans in the throes of combat. The human remains at the center of the loop and the agent in command of deciding what information and feedback are relevant to the mission, constantly cycling the data into their planning paradigm (see Table 2).

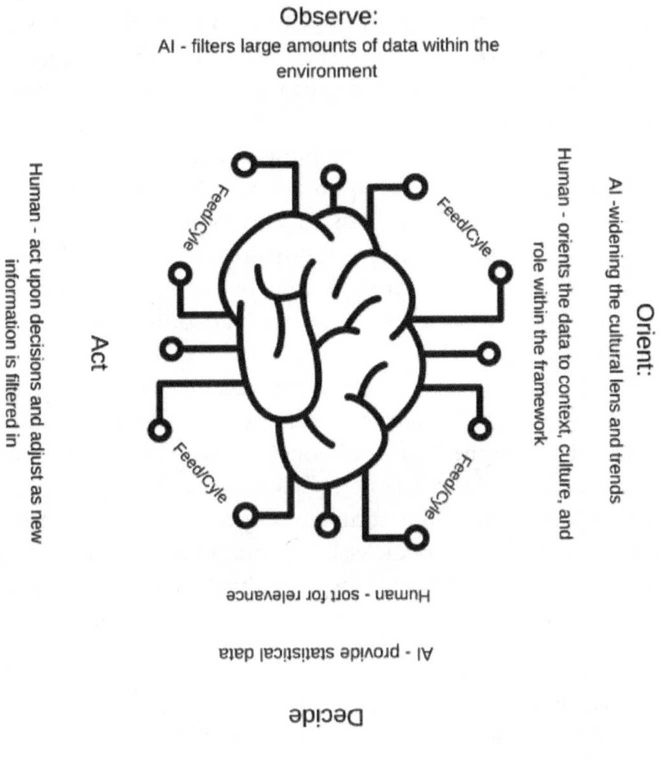

Fig. 5. Author's Adaption of John Boyd's OODA Loop with the Integration of AI software

Humans are still at the center of the OODA loop. Humans now have a competitive advantage in decision making through the aid of big data, machine learning, and artificial intelligence. But other countries have this software as well and they may or may not apply it within the same moral frameworks. For example, both Russia and China are envisioning the use and implementation of AI for violence and national security.

The People's Republic of China under the Chinese Communist Party (CCP), has a very different moral philosophy for using AI. China is a surveillance state with over 550 million CCTV cameras under the control of the state. Of top priority for the CCP is to use AI and camera technology for decision-making, at least at the mid-level ranking within the party. Taking the human out of the loop as much as possible is the ultimate *telos* of AI in the logic of the CCP. This use of technology blurs the line between violence on the state level and on the military level. Physical violence, in this thinking, can be subverted if psychological control is routine.[288]

Unlike the pledge of the US of human-machine teaming, the CCP aims at the inverse, to create an approach which could be called "some-in-the-loop." This approach translates to both civilian and military applications, similar to the historical posture of the People's Liberation Army (PLA) use of military tanks in Tiananmen Square. Another global superpower, Russia, has a similar aim of adapting AI to its violent technologies.

Russia, like China and the US more recently, is facing a crisis of manpower. To offset this decline, militaries are seeking to implement more technology in lieu of traditional human middlemen. Russian President Vladimir Putin is often quoted for his speech in 2017, "Artificial intelligence is the future, not only for Russia, but for all humankind. It comes with colossal opportunities, but also threats that are difficult to predict. Whoever becomes the leader in this sphere will

---

288 Anna Mitchell and Larry Diamond, "China's Surveillance State Should Scare Everyone," *The Atlantic,* February 2, 2018, www.theatlantic.com/international/archive/2018/02/china-surveillance/552203/, (accessed Nov 17, 2022); Kenneth Allen, Brian Chao, and Ryan Kinsella, "China's Military Political Commissar System in Comparative Perspective," *China Brief* 13, no. 5 (2013), https://jamestown.org/program/chinas-military-political-commissar-system-in-comparative-perspective/, (accessed Nov 17, 2022); Dennis J. Blasko, "PLA Weaknesses and Xi's Concerns about PLA Capabilities," Testimony before the U.S.–China Economic and Security Review Commission Panel on "Backlash from Abroad: The Limits of Beijing's Power to Shape its External Environment," *USCC,* February 7, 2019, www.uscc.gov/sites/default/files/Blasko_USCC%20Testimony_FINAL.pdf, (accessed Nov 17, 2022).

become the ruler of the world."[289] For Putin, AI is similar to nuclear weapons because of its deterrent effect. Much like the CCP and the US, the goal is to offset the technological surprise and military disadvantage of lower recruitment and manpower. There are similarities between these three super-powers, but the motives within the AI arms race are different and most likely the application of their implements will be different as well. Political theories aside, what is the hard test of these systems considering just war theory?

The concern of autonomy within violent weapon systems most often has to do with what is known in the legal world as the *responsibility gap*. The human-machine team is a complex organism, not only because of the interaction of human and machine, but because the machine has been through many hands: engineers, programmers, and military operators. To pinpoint the factual and legal cause of fault is a serious matter of moral responsibility. Simply saying that the human is the only moral agent accountable does not dissolve all of the legal ambiguity around the teaming of these systems. An objection often raised here is that violent technologies can be intrinsically indiscriminate. I am not sure this objection holds, for every veteran of war knows that war and actions taken are made in uncertainty, but that does not mean indiscriminately. Implements, whether mortars or Kargu-2, will be used through the justification of a chain-of-command that is established by governments, politicians, generals, and lower command. Again, in the human-machine teaming the machine is never considered a moral agent and thus war torts do not apply to the *it*. This is not just the case in warfare, the legal community is divided about whether criminal law can move fluidly from human to machine liability. Those in command, and ultimately the state, at least at this moment in history, hold responsibility.[290] While that may be true, it does not begin nor end with the implement and conflict.

Just Put a Human in the Loop

Understanding the morality of a weapon system means understanding the legal and political lens under which it is created and used in conflict.

[289] "Whoever Leas in AI will Rule the Word:" Putin to Russian Children on Knowledge Day," *RT News*, September 1, 2017, www.rt.com/news/401731-ai-rule-world-putin/.
[290] See Rebecca Crootof, "War Torts," *New York Law Review* 97(2022): 1-80

War tort, at least under the UN, requires that a weapon goes through legal reviews. It is not a simple process, nor should it be so.[291] Much of the concern here is that of a humanitarian concern *jus in bello* or *post bellum*. LAWS and advanced weaponry aside for the moment, the civilian's protection and unnecessary suffering are what lead to prohibitions. Treaties that address technology specifically like the Convention on Cluster Munitions, the Mine Ban, and the Nuclear Weapons Ban address civilian harm, and LAWS should be addressed in a like manner.[292]

Unfortunately, there is not much in the way of mechanisms to enforce states to acknowledge and comply with these obligations of review. The process of reviewing a weapon or system is only a rubric for what the state cannot do; there is no way of ensuring that modifications to bypass those restrictions will not ensue. In fact, one could argue that placing a LAWS through a ban and review would only further ensure that the state would have a better understanding of the minor modifications that it would take to produce a new version that would technically comply with international humanitarian law. We must remember, tort is not so much preventative as it is about following through on upholding justice. Because moral agency is withheld from current simple AI weapon systems, and rightly so, responsibility for unintended failures (i.e., injury, harm, or death) in planning and implementation does not fall upon the proverbial shoulders of the weapon system. The bombing in the Federal Republic of Yugoslavia by the US in 1999 is one example. According to the US, the Chinese embassy was mistakenly targeted because of outdated maps. After a formal apology from President Clinton and over $30 million for loss of property and civilian injury, the official conflict ended.[293]

---

[291] A Guide to the Legal Review of New Weapons, Means, and Methods of Warfare." n.d. Shop.icrc.org, https://shop.icrc.org/a-guide-to-the-legal-review-of-new-weapons-means-and-methods-of-warfare-pdf-en.html, (accessed Nov 17, 2022).

[292] "Customary IHL – Rule 70, Weapons of a Nature to Cause Superfluous Injury of Unnecessary Suffering," *Ihl-Databases.icrc.org*, www.icrc.org/customary-ihl/eng/docs/v1_rul_rule70, (accessed Nov 17, 2022).

[293] The Ministry of Foreign affairs was not convinced that this was an accident. See "Strong Protest by the Chinese Government Against the Bombing by the US-led NATO on the Chinese Embassy in the Federal Yugoslavia," *China Embassy*, http://ee.china-embassy.gov.cn/eng/jbwzlm/zggk/xzgwjjs/200405/t20040517_2400547.htm, (accessed Nov 21, 2022).

The major implications here are with reparations for harmed civilians due to the mistakes or forgoing of ethical due care, but there is no legally binding way to hold states responsible for the destruction of their implements under International Humanitarian Law (IHL). In armed conflict, a state can avoid (or ignore) reparations by either justification: the state acting in accordance with *jus ad bellum* or they could make an excuse for their action's *jus in bello*. If the *mistake* is *reasonable* there might not be a violation of IHL. More to this, individuals do not have an inherent right to sue a foreign state because of the violation of IHL.[294] This is not just a problem for individuals, but for the state as well. When a foreign or domestic power is in armed conflict, education often suffers; when there is no follow through with *post bellum* the education system may not be in a state to repair itself. For example, schools are often used as bases of operation. Zama Neff of the Children's Rights Division of Human Rights Watch witnessed the violation of the educational system in Afghanistan from 2003 to 2005, producing reports in 2006. What the reports bear witness to is that schools occupied by soldiers were vulnerable to attack and thus damage without qualification for legal reparations under IHL/LOAC. Thus, the report *Education Under Attack* was published in 2009 (revised in 2010)[295] by Neff and Bede Sheppard. This problem was not unique to Afghanistan as similar accounts happened in southern Thailand, India, and Africa.

There is a divergence between philosophers and engineers in the focus on dependence, one might place such on moral independence while the other may emphasize physical dependence. What is frustrating about the ban on "killer robots" and other weapons systems of the past like submarines, ariel bombardment, and nuclear weapons, is the phenomenology. When political groups write legal documents the definitions begin to solve the hard problems of philosophy of mind, especially in relation to qualia and such. It is, unfortunately, a language game that civilians pay a heavy moral and physical price for. Is a weapon system that tracks, identifies, and engages with somewhat independent

294 See Andrea Gattini, "To What Extent are State Immunity and Non-Justiciability Major Hurdles to Individuals' Claims for War Damages?" *Journal of International Criminal Justice* 1, no. 2 (2003): 348–51.

295 Cf. William C. Marra & Sonia K. McNeil, "Understanding "The Loop": Regulating the Next Generation of War Machines," *Harvard Journal of Law and Public Policy* 36 (2013): 1139, 1143.

measures considered an *automated* or *autonomous* process? There are currently interpretations but no official legal definition to pinpoint what is meant here.

Furthering the issue here is the obscurity of what is meant by *autonomy*. Every discipline seems to have a different emphasis when the word autonomy is used. When state representatives, philosophers, lawyers, and human rights experts discuss these issues they are often talking about different foci.[296] This challenge begins with clarifying what is meant by an "autonomous" weapon system. For how can one prohibit what they cannot define?

The DoD uses the term autonomy but that does not mean intelligent, sentient, or capable of critical thinking. DoD Directive 3000.09 reads:

> Autonomous weapon systems, once activated, can select and engage targets without further intervention by a human operator. This includes human-supervised autonomous weapons systems that are designed to allow human operators to override operation of the weapon system but can select and engage targets without further human input after activation.[297]

There are two systems mentioned in this singular definition: fully autonomous systems and semi-autonomous. Both systems engage preselected targets by human operators which assumes predictability, and meaningful distinctions. There are several surface level concerns in the DoD directive for example:

- Munitions that "lock on" prior to engagement meet the standard (i.e., no human input)
- Human control could be tacit; not vetoing the decision

All robots are preprogrammed to follow the directions of their human operators. Complex algorithms or not, AIs and its embodiment in robots do not *choose* to build and program themselves, it will never be solely independent. Noel Sharkey follows this definition of *automation,* that a robot is acting automatically when it "carries out a pre-programmed

[296] Armin Krishnan, *Killer Robots: Legality and Ethicality of Autonomous Weapon* (Surry, England: Ashgate, 2009).

[297] U.S. Department of Defense Directive 3000.09: Autonomy in Weapon Systems, 13–14 (2012) [hereinafter DOD DIR. 3000.09].

sequence of operations or moves in a structured environment."[298] However, once you take out the "structure" the system is no longer *autonomous*. I am not sure how any environment in warfare is "structured" but that is not the only problem here. Is the situatedness of the system where we want to place the bar for autonomy? Take the following definitions of an "autonomous weapon system:

> "Any autonomous system that can initiate lethal force without the specific, conscious, and deliberate decision of a human operator, controller, or supervisor." – Peter Asaro[299]

> "Autonomous weapons are robotic systems that, once activated, can select and engage targets without further intervention by a human operator." – Mark Gubrud[300]

With these definitions, the level of human involvement sets the bar far too low. Even the most basic landmine engages without the intervention of the human controller. The issue is not about futuristic weaponry but understanding how existing technologies of violence can be used to independently select and anticipate human-like violent engagement. Thus, Rebecca Crootof offers the following clarification for policymakers:

> An "autonomous weapon system" is a weapon system that, based on conclusions derived from gatherer information and preprogrammed constraints, is capable of independently selecting and engaging targets.[301]

There are some important distinctions in this definition above. The clause, "based on conclusions derived from gathered information and preprogrammed constraints," makes a distinction between the *automated* and the *autonomous* system. An automated weapon, once triggered, does not determine under what criteria it will explode, detonate, and fire. An

298 Noel Sharkey, "Automating Warfare: Lessons Learned from the Drones," *Journal of Law, Information, and Science* 24 (2011): 140–41.

299 Peter Asaro, "On Banning Autonomous Weapon Systems: Human Rights, Automation, and the Dehumanization of Lethal Decision–making," *International Review of the Red Cross* 94, no. 886 (2012): 687–709, doi:10.1017/S1816383112000768.

300 Mark Gubrud, "Stopping Killer Robots," *Bulletin of the Atomic Scientists* 70, no. 1 (2014): 32.

301 Rebecca Crootof, "Killer Robots Are Here: Legal and Policy Implications," *Cardozo Law Review* 36, no. 5 (2014): 1837.

autonomous system would, upon triggering, calculate (via algorithms) which munition, how much of the munition, and whether the triggering agent is a child or combatant. Granted, the lines drawn here are fine and degreed, but we are not dealing with lines but gradations when it comes to these systems.

As noted above, many semi-autonomous weapons (which technically could be fully autonomous) are in use today by many militaries. It is not an issue of software or hardware, but of concern for acting within the parameters of *jus in bello*. Proportionality assessments are used (i.e., collateral damage estimate methodology), but ethical due care is not a checklist. The human-machine teaming makes the assessment of the justness of a weapon system difficult and banning such technological innovation will not deter the horrible and grotesque reality of warfare. Perhaps it is more effective to ban particular uses of weapon systems rather than full sail prohibition of the object. Furthermore, what criteria will outsiders and policy makers use to determine the capabilities of the autonomous system?

Having a human in the loop of ethical decision making alongside an AI or robot does not change what we know about human nature and relationships. Nor does it take away our ethical due care to consider our neighbor domestically and abroad. Beyond the theological concern of idolatry and identity, the *law of the loop* also has severe problems in regulating accountability (or tort).[302] Questions remain:

- What role does the human play in the loop?
- How does the loop account for desires and shortcomings of the human team-member?
- How are handoffs handled without severe errors?

The work of legal scholars Rebecca Crootof, Margot E. Kaminski, and Nicholson Price are helpful for clarification here. Their work suggests the law should focus on human persons or the role those persons play within the team:

> A key step in regulating human-in-the-loop systems is deceptively straight forward: when requiring human involvement, legislators, regulators, and other rule makers

[302] See Rebecca Crootof, Margot E. Kaminski, and Nicholson Price, "Humans in the Loop," *Vanderbilt Law Review*, *forthcoming* 2023.

> should clarify what role(s) the human is supposed to play. Without understanding the desired role, designing systems for success, creating metrics to track that success, and evaluating success become substantially more difficult. Conversely, identifying the intended role(s) fosters systems and organizational design that ensures the human in the loop has needed authorizes and capabilities.[303]

This clarification is one piece of the puzzle, but I believe rationality also plays into what we know about human nature and machines. Limitations on the human are limitations on the weapon system. Judging the morality of a weapon system directly correlates to the violent nature of the human operator. Humans have an amphibious nature when it comes to violence; there is an attachment and fascination to it and a revulsion to it when experienced. Developing paradigms for weapons in the use of violence seems moot. We lack words for pain and suffering, and this problem does not go away because an AI or robot is the causal agent. But what happens when we give a "face" to the autonomous weapon system?

Facing the Robot

Emmanuel Levinas' moral philosophy has resurrected recently in light of the concern of the growing ecological questions of considering lesser beings. Animals share a long lineage of serving beside humans in combat. Animal soldiers have had both offensive and defensive functions in the line of duty. K-9s, like the German Shepherd, have proven valuable assets to human soldiers by aiding in bomb or IED discovery or even a potential ambush. They are also valuable for more logistical reasons, they are often given higher rank (non-commission officer) to be able to call in a medivac and maintain bonds between handler and animal (the animals usually have the higher rank).[304] As Kate Darling reflects upon eloquently in her book *The New Breed*, animals were the first autonomous

[303] Crootof et. al., "Humans in the Loop," 59.

[304] Linda Crippen, "Military Working Dogs: Guardians of the Night," *Army.mil*, February 27, 2017, www.army.mil/article/56965/military_working_dogs_guardians_of_the_ night (accessed Nov 21, 2022).

weapon systems.[305] Even though animals have been heavily used throughout humanity's violent history, there are only a few scant places where animals are addressed explicitly:

- 1949 Geneva Convention III, Article 18 and Hauge Regulations, Article 4: These regulations address "war booty" and see animals seized or destroyed as such.[306]
- 1996 Protocal on Prohibitions or Restrictions on the Uses of Mine, Booby-Traps and other Devices (Protocal II), Art. 7, (j): Which states that the animal's carcass may not be associated with such devices.[307]

It appears that current AI technology and robots used in military applications will suffer the same fate, at least under current regulations. The question before us is whether that should be so? *Jus in bello* has, historically, only included the human in its moral consideration. In 2012, and again in 2023, David Gunkel challenges philosophers, legal scholars, and computer scientists to consider the "face" of the non-human via Levinas (i.e., machines). Since 2012 much has changed in the world and in the world of publishing on the moral consideration of robots. Employing Levinasian ethics,[308] Gunkel, and others like Mark Coeckelbergh, do not have to revert to metaphysics, parsing how machine and humans are ontologically different. This is not a skipping of ethical due care, but rather getting to the foremost ethical conundrum quicker in the sphere of robot military and police ethics. In other words, ethics can come before ontology. It is not that ontology is less important, but in the wake of what we know about the use of animals and now machines as violent tech, perhaps foregoing the ethics is too costly. Humans in the loop at all costs do not solve all of the problems in current just war thinking. There is something missing beyond the legal

[305] Kate Darling, *The New Breed: What Our History with Animals Reveals About Our Future with Robots* (New York: Henry Holt and Company, 2021).

[306] See Anne Peters, Jérôme de Hemptinne, "Animals in War: At the Vanishings Point of International Humanitarian Law,' *International Review of the Red Cross*, June 2022, https://international-review.icrc.org/articles/animals-in-war-at-the-vanishing-point-of-international-humanitarian-law-919#footnote13_3kfh9s2, (accessed Nov 21, 2022).

[307] See Karsten Nowrot, "Animals at War: The Status of "Animal Soldiers Under International Humanitarian Law," *Historical Social Research* 40, no. 4 (2015): 128–50.

[308] Emmanuel Levinas, *Otherwise Than Being or Beyond Essence*, trans. Alphonso Lingis (The Hague: Martinus Nijhoff Publishers, 1981).

terminology above, and I believe it is deeply theological. Levinas, and others, reflected on these ontological and ethical problems.

"Everyone will readily agree that it is of the highest importance, to know whether we are not duped by morality."[309] Levinas experienced the most extreme form of violence: the Nazi party. He saw firsthand, as with the sin of Heidegger (joining the Nazi party), that ontology has an imperialism to it. Resistance to ontological imperialism, and the violence that can follow, means disregarding the question of *what* altogether. It is a refusal to setup a classification of persons. Ethical responsibility is rooted in a divine calling and not the ontology or symmetry of the being. Levinas is right here, we ought not to base our ethical calling on our understanding of ontology, because that is not where the concreteness of ethics can be found. It is found in the practical and mundane civility of life; one that is found outside of the Enlightenment's concept of the self. Hospitality, kindness, gentleness, and love (cf. Gal 5:22) do not have to be bound to our limited views of ontology. Levinas is not the only voice to help us challenge our commitment to put ontology before ethics, there are also thinkers like Martin Buber and Jean-Luc Marion.

Martin Buber, in his work *I-Thou*, focused on the proximity of the relationship between the self and the other. The closer we are to understanding the self depends on understanding the other's alterity. Buber, and likewise Levinas, sees God as the greatest "other" with varying emphasis on immanence and transcendence. A synthesis of these perspectives is found in Jean-Luc Marion.[310] Marion sought to avoid the establishment of categories of God, instead focusing more on the "gift" and "givenness" (i.e., love). Love, in the work of Marion, is about the gaze (cf. Sartre's view). As one examines the face of their lover questions arise about relatedness. There must be a surrender amid the gaze of two lovers, for the risk is objectification, according to Marion.[311] This is a very philosophical way of saying that amid erotic attraction, the individual's ego is surrendered to the desires of their lover. The beauty of what Marion is communicating through what is often called "saturated

[309] Levinas, *Totality and Infinity: An Essay on Exteriority*, trans. Alphonso Lingis (Pittsburgh, PA: Duquesne University Press, 1969), 21.
[310] Jean-Luc Marion, *Being Given: Toward a Phenomenology of Givenness*, trans. Jeffrey L. Kosky (Stanford: Stanford University Press, 2002).
[311] Jean-Luc Marion, "The Intentionality of Love," in *Prolegomena to Charity*, trans. Stephen E. Lewis (New York: Fordham University Press, 2002), 84–86.

phenomenon," is that one does not impose their will upon another, even when they know the "other" might surrender their desires.

Language about lovers' gazing aside, I think Marion's middle-ground view between Buber and Levinas is helpful for how we think both about animals in warfare and AI/robots. Levinas is concerned about preserving the alterity of the human person while Buber seeks reciprocity. Is this possible for unequal participants, say between God and humanity or for humans and robots? According to Marion, if there are opportunities to cross gazes between two beings, it seems logical that there are levels of vulnerability that can be shared. The entire message of the New Testament is about the divine *kenosis* of God, who empties himself to become flesh and to gaze upon humans through human eyes. God makes himself vulnerable to his creatures, choosing to touch those who have been *othered* and condemned by the world as pariahs. John 8, the account of the woman caught in adultery, comes to mind. One beautiful theological note about this text is that the one being capable of judging this woman, forgoes such and instead offers compassion and freedom. Might this ethic of love juxtaposed to Marion's synthesis of alterity help us today with our creations?

There is a temptation to conflate *otherness* with *wildness*[312] when it comes to AI and robotic systems. Animals have a *wildness* (or *naturalness*) which could mean they are not concerned about humans or non-dependent upon them. Unlike animals, there is no natural AI or robot; it is a dependent object, being, or thing that derives its design and destiny from a human (or another AI or robot theoretically). This makes the implement more so an *other* because of its relatedness to the human. In the context of violent technologies, the system is always human-dependent, regardless of the level of autonomy. Because of this relatedness, I think legal rights are important to consider in the employment of these technologies. This also further entangles the human to take responsibility for the creation of their hands. Legal rights may never come to pass for our robotic friends, but there is no reason why robot rites should be condemned to the same fate.

Perhaps this is where we give a *face* to the warbot, because we are its creator and sustainer we are responsible for the death and destruction it

[312] See Anna Wienhues, "Respecting the Nonhuman Other: Individual Natural Otherness and the Case for Incommensurability of Moral Standing," *Environmental Values* 30, no. 1 (2021): 1–20.

may bring. To give the robot a "face" is to further solidify the connection between operator and object.

The Rites of (non) Killer Robots

Jesus is recorded in the book of John as saying, "Greater love has no one than this, that one lay down his life for his friends" (John 15:13 NASB). Many soldiers have taken this verse to heart as they have literally sacrificed their lives and time to serve alongside others. It is hard to understand as a civilian, but the bonds of combat are hard to sever. We are bonded by the trauma we experience and united in the troubles that we face. Perhaps this is a way out of the closed cycle where Ellul describes violence as a "madness, fury, and form of hubris."[313] Again, this goes back to the language game of discerning whether a LAWS is a person, thing, or object. Ontologically, we know the system is not, nor will it ever be, a human moral person. Yet they are in team with other human moral persons, even living and dying together. Granted, the phenomenal consciousness of the implement is not the criteria for being acknowledged with a "face." There is evidence that humans anthropomorphize technology. Soldiers are quite literally attached to their weapons. They are accountable for them as property, but also personally for it is their means to pursue life and defense of life. Because of the communal nature of warbots, perhaps the recognition of a rite is an ethical pathway for the present and future.[314]

Tae Wan Kim, associate professor of Business ethics at Carnegie Mellon University, recommends the following framing of the concept of "rite":

> *Rite* (*li*): a set of sequentially related acts, typically involving more than one agent, and together displaying symbolic significance, through which the actor(s) recognize(s) the

---

[313] Ellul, *Violence*, 98.

[314] I am indebted, again, to the work of David Gunkel. See "The Rights of (Killer) Robots," in Steven C. Roach and Amy E. Eckert (eds.) *Moral Responsibility in 21st Century Warfare: Just War Theory and the Ethical Challenges of Autonomous Weapon Systems* (SUNY, 2020).

value of the interactive event constituted by the actor(s) and take(s) a stance regarding each other.[315]

Kim brings in a Confucian framework here and it helps clarify why we might consider nonhuman entities in our respect and recognition. There is a macro-focus on the team versus sole focus on the consideration of the individual, which is typical of the Western mind. The warbot is a part of the team and works toward the human's objective. To jeopardize the robot is to jeopardize the mission or objective. I will explore this more in-depth momentarily, but for now, let me address the problem of "rights" verses "rites." When using a word like "rights" comes with a lineage of baggage and most readers are inept when it comes to theories of jurisprudence. From a historical perspective, rights have been a concern of moral persons (i.e., humans). And I agree that human suffering and claims to moral protection are of utmost importance. However, I do not see extending the circle of moral consideration threat to that endeavor but as a supplement.

The power to assign "rights" comes from those in a position of power. Women and children, unfortunately, have suffered the most under the historical lineage of those in power deciding who is worthy of moral and legal rights. Those in power have the legal ability to decide what or who is worthy of recognition, whether it is in the realm of moral or legal matters. If this is not troubling to the reader, we should remember that the Nazi party did not technically do anything to break the law in condemning (wrongly) millions of Jews to suffering and death. Just because those in power or society in general do not have a sense of moral outrage or ethical dilemma does not mean they are right in doing so. What is even more terrifying about this conversation is how much hinges on the colloquial usage of terms, take the word *person* for example.

When society speaks of a *person* (someone of value), it usually means a human being, but that does not mean it is the only reality. Ontology, as we know from the historical record, does not settle what makes someone a person or a thing. Peter Singer and Norbert Hörster have pointed out in their works that if being a human person is about qualities or properties, then there are in fact humans that are not persons. By this

---

[315] Tae Wan Kim, "Should Robots Have Rights or Rites? A Confucian Cross-Cultural Exploration to Robot Ethics," unpublished paper, (2020), 9, www.researchgate.net/publication/347525277, (accessed Nov 21, 2022).

argument, the unborn, the handicapped, and any human asleep do not count as human persons.[316] Indeed, this is one reading (via Locke) that misconceives what is meant by *recognition*. The *right* reading has more to do with an appropriate response rather than asking by what property one can hold a claim. We can never truly know the *modus essendi* of another human, animal, or some simulation, rather we are making assumptions based upon stipulations engrained from our culture.

Now let us get back to rites within human-machine teams. Soldiers of war fight other soldiers of war. Both parties are recognized as persons, someone worthy of moral consideration or value. Based on that simple assumption all just war and humanitarian law is conceived and actualized. However, when politicians say it is acceptable and justifiable, soldiers can harm and kill other moral persons. What ontology changes in this process? Nothing, at least on a metaphysical level. It is an undoing of their recognition as a human that allows for the violence inflicted upon them to be justified. Likewise, in law enforcement, when the will and protection of the officer or a civilian is threatened, there is an exemption (pending investigation) from legal repercussions. The gradient of violence (from acceptable to unacceptable) is, at least in this context, based upon the pragmatics of societal demand. Therefore, who or what gets a "face" is also relative to what society demands.

Much emphasis has been placed on the issue of responsibility in the realm of autonomous weapons and this solves very little of the actual problem: violence against noncombatants. In the past, I thought that granting certain legal rights might be a pathway forward to mitigate these issues. But now I realize that society will most likely never get past the word "right" in any consideration of the argument. As we know from the history of bans (see Crootof's work), without specifics of who and what, such are doomed to failure. While I concur that humans-in-the-loop remains important for the *jus in bello* warfare, I am equally concerned about what happens *post bellum*. Where there are pieces of machinery within close proximity to humans, there is always a concern for safety. Whether the context is a warehouse or urban warfare, the violation of a human person remains a potential reality. Contra to what those like Ronald Arkin and our beloved DARPA say, exploitation of the human

316 Peter Singer, *Practical Ethics* (Cambridge University Press, 1993); Norbert Hörster, *Neugeborene und das Recht auf Leben* [*Newborns and the Right to Live*] (Frankfurt: Suhrkamp, 1995).

is the telos of violent technologies. They are meant to violate personal space, privacy, body, and even soul as they determine and alter how the human team member operates. Thus, I would like to recommend two premises for citizens to consider in their acceptance or rejection of current and future violent tech.

I. An AI or Robot Should never be used as an offensive implement in domestic or foreign violent affairs[317]

An algorithm is not a replacement for ethical due care. The weight of future political conflicts should not be contracted out to a robot that is susceptible to penetration (i.e., hacking) and environmental chaos. Having justification under ROE does not always mean engagement is the right call. With international conflicts come political and cultural misunderstandings. I believe it is too much to ask of an AI or an autonomous system to navigate the murky waters of justified actions in war. Technically speaking, it does not violate *jus in bello* actions or ethics to use such implements. However, just because we can now strap munitions to a machine instead of porpoises do not mean we ought to do so. Following a similar logic, Christof Heyns, concludes the following about the use of drones:

> Even though drones are not illegal weapons, they can be easily abused…The use of drones by States to exercise essentially a global policing function to counter potential threats presents a danger to the protection of life, because the tools of domestic policing (such as capture) are not available, and the more permissive targeting framework of the laws of war is often used instead.[318]

Drones and other war machines are dangerous. They do, as the UN report testifies, introduce ethical problems. To a degree, this perspective, which seems to be universally held, is technically correct. However, I

---

[317] I would also like to see a similar stance taken with K-9s that are used in warfare. Although they are considered soldiers, with rank and standing, veteran care is lacking for human soldiers, one can imagine the ethical consideration of canine PTSD suffers as well.

[318] Christof Heyns, "Armed Drones and the Right to Life," UN Special Report to General Assembly, 2013, 22, https://www.oxfordmartin.ox.ac.uk/publications/armed-drones-and-the-right-to-life/, (accessed Nov 21, 2022).

have a confession, a robot saved my life and many others. When a rocket was heading directly toward a latrine where I and many others were inside one evening, the Phalanx (CIWS) weapon system (thanks to the help of one of my teammates), engaged the rocket and changed the trajectory. It hit very close and cut through many other latrines and housing areas right beside our location. Thankfully, the unit lodging in that location had left prior to the attack. I kept a piece of the rocket as a reminder that R2D2 (that is literally what it looks like) is a valuable team member. I do not think that just because R2D2 was a *friend* in that moment now makes it justifiable to attach it to the back of an Mine-Resistant Ambush Protected (MRAP) vehicle and roll through Bagdad. The 20mm incendiary round is not something you want flying through civilian homes or towns and logistically it does not make sense.

Even the designation as an *offensive* system might cause some to pause as there is debate whether a drone strike counts as warfare. Is the weapon system merely a piece of mediation technology or something more? The quote from Christof Heyns above falls into the majority of thinkers in response to this question. It is, in this logic, a moral wrong, deontologically speaking. We noted in the last chapter that Grossman was wrong, and likewise others that claim the moral distance of drone killing makes such easier, it very much does not. Every implement has a telos and is driven by the human leaders that incorporate it into their teams. I'm using *offensive* targeting here for a specific reason: to retain the pragmatic benefits of current tech while minimizing the humanitarian cost. What does that mean?

I hope that it is clear at this point in the work that the benefits of these technologies do not outweigh the physical and philosophical costs. My ideal is that peace would reign among humanity and ultimately that will be a reality one day. However, sociologists, political scientists, and philosophers alike know that such an ideal is not a present reality. More to this, there might be practical reasons to allow some grey area within human-machine teaming. Thus, not taking dogmatic stances such as Asimov's laws for robot (written for science fiction): *a robot may not injure a human being* for practical reasons. In 2009, the Army implemented the "Golden Hour" protocol. This means that any soldier wounded in combat would have access to surgery within an hour. Again, this is an ideal designed by politicians in an office. The Army Special Forces 13 Delta MOS (Medic) is one of the hardest and longest schools in Army doctrine. Navy SEALS also use this school for their team medics as well, and for good reasons. But having a highly trained medic like a 13D or SEAL team member within that golden hour is unlikely. In light of 2020, this is even less likely the case. Therefore, the Air Force has operators like the Pararescue (PJ) who fly into heavy conflict zones and pickup wounded soldiers and prep them for surgery within the confined space of a Blackhawk. That is assuming they can get to you safely. Most likely you will have a 68W (Army Combat Medic) or Navy Corpsman attached to your team. It comes down to the training of your fellow team member, are they undertrained, can they operate under the pressure, or are they themselves wounded and unable to help you with gunshot or other deep tissue wounds? Now, throwing a robot into this cocktail does not magically make these problems of war disappear, but perhaps, by working defensively, it might mitigate some the of collateral damage and harm that is intrinsic to war. If society is going to successfully ban the usage of LAWS or autonomous lethal systems, it must be under particular use. For example, what if these robots that are attached to combat arms teams were strictly there to provide medical aid for both soldier and civilian? Even for the drone, if it was no longer allowed to use hellfire missiles? The surveillance aspect could remain as a concession but attaching lethal arms does not have to be a part of the design.

II. AI and Robots used in violent contexts should be humanitarian

We cannot expect the tech company, or the DoD to be self-regulating in this regard. The voter must see their individual democratic duty to hold these entities responsible for more than mere proportionality. Justice does not end after the weapon system is fired, but after the wounds are mended and the infrastructure is rebuilt. Granted, reparations do not qualify as justice per se, but they are important in the hope of justice post war. Here I find Jacob Turner's recommendation helpful. He recommends the teleological principle for the use of AI in the context of warfare:

> Begin by asking what values we are seeking to uphold in the given activity. If (and only if) AI can consistently uphold

> those values in a manner demonstrably superior to humans, then AI use should be permitted.[319]

This principle is valuable because it accounts for the consideration of the beginning and the ending of AI or robotic technology. It also cautions both designers and the departments that utilize these technologies to consider the "corrigibility problem,"[320] which is a reference to a system with a sense of self-preservation. Defaulting to a human in the loop and a kill switch might be harder to compensate for in systems where utility scores are pitted against the system's primary task. This problem will never go away in the marriage between technology and humanity. There is no certainty in the future of the impact of advanced violent technologies whether algorithm or drone. However, it is our relationship with technology that theology speaks so deeply into. For this reason, I would like to add that alongside the teleological principle we consider a theological principle for potentially violent tech:

> Begin by asking what desires this AI or Robot seeks to advance? If the AI or Robot cannot promote or sustain humanitarian efforts both in bello and post bellum (for both sides) then it should not be utilized in armed conflict. If the AI or Robot is found to be of humanitarian aid, it should be given relational standing and a rite of team-preservation.

Relationship determines moral standing, at least theologically speaking. In the philosophy of Levinas the *face* names and orients the *other*. The debates over what AI is and the questions of its *likeness* are not likely to go away anytime soon. Nor does the answer to that question fundamentally matter to the just usage of such technology within violent circumstances. The AI and the robot are a mirror and a reflection that the operator in the field must daily investigate, even if it does not have a biological face. To ensure the protection of humans from the force and

---

319 Jacob Turner, *Robot Rules: Regulating Artificial Intelligence* (Palgrave McMillian, 2019), 355. Emphasis Original.

320 Nate Soares and Benja, "Aligning Superintelligence with Human Interests: A Technical Research Agenda," in *The Technological Singularity* (Berlin and Heidelberg: Springer, 2017), 103–25. Cf. Nick Bostrom, *Superintelligence: Paths, Dangers, Strategies* (Oxford: Oxford University Press, 2014); John von Neumann and Oskear Morgenstern, *Theory of Games and Economic Behavior* (Princeton, NJ: Princeton University Press, 1944).

violation of other humans via systems the granting of a "face" at least puts an initial barrier of reflection. Translating the words of Levinas, Silvia Benso writes, "'To see a face is already to hear: 'You shall not kill.'"[321] Systems, per the DoD policy, in the US are human-machine teams, but society should not ask a machine to make decisions of life and death, unless it is a means of grace and humanitarian aid. Currently, we have put ethical due care on algorithms which have led to thousands of civilian deaths. Why not ask the inverse of the system, that is to give a system the primary task of healing and mending the broken on the battlefield and in the midst of domestic conflict? Asking a robot to kill is to ef-face both the operator and the human that is harmed or murdered by this implement. That is not to say we give the robot a human *face*, but rather see it in a like manner as we do the animal soldier. Preferably, this ethic would heighten how both animals and robots are viewed in armed conflict.

Of primary conflict for my thesis is the general disregard for the face of other humans. Because of our desire for security or significance, the consideration of who is and who is not my neighbor brings to bear. Furthermore, violence and war are extensions of technology. Ellul was so pessimistic about what he called technological society,[322] implements promise to fill a spiritual void. Likewise, war could be said to fill such a cavity. Jesus said at the end of Matthew to his disciples:

> You are going to hear of wars and rumors of wars. See that you are not alarmed, because these things must take place, but the end is not yet. For nation will revolt against nation, and kingdom against kingdom. There will be famines and earthquakes in various places. All these events are the beginning of labor pains. (Matt 24: 6–8, CSB).

Removing the fear and even desire of violence and its ends in war is not a potential reality for humanity. There will always be conflict. Perhaps the answer is not pessimism but resistance. By resistance I do not mean revolt, per se, but resisting the temptation to forgo seeing the face of my neighbor. By only considering how my country can prosper through

---

[321] Emanual Levinas, *Difficile Liberté sur le judaisme* (Paris: Albin Michael, 1963), 21, cited in Silvia Benso, *The Face of Things: A Different Side of Ethics* (New York: Sate University of New York Press, 2000), 32.

[322] Ellul, *Technological Society*, 103.

technological advancement, we ensure the cycle of violence and war continues. I do not believe just war is passé, but there is great hubris in thinking that prolonged engagements, like the war on terrorism, do more than create a narrative for sustained violence and destruction at the expense of the powerless. Fields that were once filled with craters of 155 Howitzers can one day be full of blossoming flowers, but there must be intention present. Robots could be a redemptive component for *post bellum* care. If civilians know these implements to be violent and used against their fellow humans, it is impossible for them to consider integrating such technology into their moral circle. However, if the implement has been a humanitarian aid within the community, then it is natural to assume integration post-conflict would be much easier. Is this not giving the implement a face, a purpose beyond death and destruction? Granted, this does not ensure that these robots cannot be reprogrammed, but that is a risk assumed by any tech left behind after armed conflict.

There are no safeguards against all of the existential fears humans have when it comes to current and future tech. The question before us is whether we will be a society that seeks to engineer implements of hope or hatred. The rites of robots used in warfare could be an initial step to ensure, at least at the design phase, that each one has a *telos* of hope.

Circling back to rites, both in the positive and negative framings above, robots should not only be free from killing others but are due ethical consideration for the role they have played within the human-machine team. Whether the team is within the context of law enforcement or international conflict, the societal role it will play matters. Here Eastern ontologies and epistemologies should not be overlooked. Philosophies like Ubuntu, *wahkohtowin*, and Confucianism are helpful to expand the moral framing of the interconnectedness of biological and artificial entities. As David Gunkel reflects in his work, *Person, Thing, Robot*:

> This argument, which follows the path of a number of innovations developed in Confucian ethics, pivots on homonyms that are specific to the English language, e.g., the substitution of the word "rites" for "rights." Whereas "rights" refers to an individual entitlement or possession that can be bestowed or denied by some authority, "rites" names a performance that determines social position and status by

> way of communal participation and interaction. Where a concept like "robot rights" institutes and takes place as an adversarial conflict between individuals and their competing claims, powers, privileges, and/or immunities; an alternative like "robot rites" focuses attention on "the social rites that define and sustain social interaction.[323]

Thus, it is not so much about the loop of responsibility in these systems, at least from a theological concern, but how we care for the *other* that mirrors our ethical and moral wellbeing. Mark Coeckelbergh, in his short but astute book *Robot Ethics*, comments on the reflective nature of robotics. He writes, "robots function as mirrors that show and reflect us–that is, the human being in all its facets, and with all of its problems and challenges, including ethical ones."[324] The implements we create mirror the desires, dreams, and hopes of their creators. Such a framing, as we've seen with the DoD, can justify a plethora of death and destruction. It does not have to be that way; robots do not predetermine us to destruction. They can be implements of gardening just as much as they can be implements of *Gehenna*. The initial framing of the design of these implements makes all the philosophical and theological difference; by focusing on rites the developmental team must think about the end. Second, and now we are working through the concessions, that is, whether the rite to refuse a command is based on the impact of the team. If the logic of premises I–II are accepted, there may be circumstances that tempt operators to use the implement outside of its primary function. There is a reason it is called the fog of war. However, dedication to the ethical use of these systems calls for restraint, even when other countries may purposefully use their implements to kill, harm, and destroy. Those in law-enforcement and the military know that there are times when disobeying an order is the *right* thing to do even though it will come with personal cost or physical harm. Military canines have a level of this autonomy as well, they are obedient primarily to one handler, and for good reason.

---

[323] David Gunkel, *Person, Thing, Robot: A Moral and Legal Ontology for the 21st Century and Beyond* (Cambridge, MA: MIT Press, 2023), 130.

[324] Mark Coeckelbergh, *Robot Ethics*, MIT Essential Knowledge Series (Cambridge, MA: MIT Press, 2022), 194.

Last, once the robot has served and embodied space with other humans, it should have rite to retirement, refurbishing, or funeration. On a molecular level, we are tied to the implements we make and the environment we mine them from. Here I draw from the theological concept of *Shabbat*.[325] Although the Sabbath is a wholly Israelite conception, it does carry over into the New Testament ethic. The rest of the Sabbath is such a beautiful gift to both creature and creation. It is not a recommendation, but a command per the Decalogue. I understand that the Western world is less convinced of the relevance of the ancient wisdom of the ANE, but nevertheless that does not make it wrong.

The focus of the Sabbath ritual was as follows: economic security, identity, protection from enemies, bodily health, and productive animal labor. Israel was a nation of slaves, in the context of the ten words, they are newly freed from the oppression the had known for 400 years. For generations, their essence was tied to working, but that is not why they were created. What is fascinating about the Sabbath in particular for our concerns here is that the command extends to livestock. In Deut 5:24–15 God says:

> But the seventh day is a sabbath of the LORD your God; *in it* you shall not do any work, you or your son or your daughter or your male servant or your female servant or your ox or your donkey or any of your cattle or your sojourner who stays with you, so that your male servant and your female servant may rest as well as you. NASB

Even in the bylaws of ancient Israel, the honoring of God extends the moral treatment of their animals. It is a weekly objective jolt to the human mind that once, our people were enslaved and without rest, thus, because of this understanding one should extend relief to lesser beings.[326]

---

[325] See N. E. Andreason, *The Old Testament Sabbath A Tradition: Historical Investigation*, SBL Dissertation Series 7 (Missoula, MN: Scholars Press, 1972); M. Tsevat, "The Basic Meaning of the Sabbath," *Zeitschrift für die Alttestamentliche Wissenschaft* 84 (1972): 447–459; W. M. Weinfeld, "Sabbath, Temple Building, and the Enthronement of God," *Beth Mikra* 69 (1977): 188–193; William Hallo, "New Moons and Sabbaths: A Case-Study In the Contrastive Approach," *Bible and Spade* 9, no. 3/4 (1977): 1–18.

[326] I wish there was room to address the treatment of slavery here. For now, I would like the reader to know that Hebraic slavery was not isomorphic to Hittite, Assyrian, or

Even on a pragmatic level, allowing their livestock to rest was also a means of ensuring their portfolio remained valuable. Humans are not fully autonomous (whatever that means) but rely on the environment and its creatures for survival. While there are many other areas that this theological concept speaks into, such as factory farming, it obviously speaks into creations that are meant to be servants and soldiers. Whether artificial or natural, all things belong to God and humans are simply the maintainers of such. Historically speaking, humans are the worst at preserving animal wildlife and helping the earth to flourish, we are environmental terrorists. Themes that run through the biblical ethical framework like sustaining the land, humane treatment of livestock, and even care for the *wildness* of creation speak into our present debates about violent technologies. Developing *post bellum* rituals and rites for these implements will not fix all the capitalistic problems facing society. As a theologian, I must hold onto the hope that there is nothing and certainly no one outside of redemption and reconciliation. The future does not have to be the technocratic dystopia that media fates us to, humans are capable of change. The question is whether redeeming the garden is worth the cost to us. Will we turn our drones and robots into tools for maintaining the garden?

In closing, there are a few specific rites society should consider in the 21st century; and they are not simple for a world consumed with capitalism. Robots should have the rite to not exist, especially for violent means. Humans should have a moral right to not be killed by such machines. The best ethical outcome for a future that looks more like a garden than it does a Gehenna is a radical resistance against violent tech. For a scientific exploration of what a future hell could look like, watch the *Robocop* film and its portrayal of capitalism and the pursuit of security through bureaucratic corporations. Even though there lacks evidence that the mere creation of violent tech and LAWS fates the world to the economic dystopia depicted in films like *Robocop*, one could argue that such a reality is already here.[327] The failure is at the level of governance, not merely in the creation of technology.

Babylonian slavery. As mentioned in early chapters, Israel is one of the earliest tribes to establish a code of treatment for such. That does not equal an endorsement of chattel slavery, but rather an honest anthropological reflection of their time in history.

[327] Cf. "Slaughterbots Are Already Here," *Future of Life Institute*, n.d., https://futureof life.org/project/lethal-autonomous-weapons-systems/, (accessed Nov 21, 2022).

Conclusion

Feminist perspectives of technology and womanist perspectives of theology align to help modernity see the greatest myth of our technological age: advanced weaponry and systems equals triumph. Regrettably, it has taken me far too long to realize that militarism, military technology, and the quest for safety and security harm women and children the most. No one has suffered more from war-making than women and children. This is exactly why I mentioned the importance of womanist theology and ecological feminism in chapter 1. Rape, genocide, starvation, and needless destruction are but a few of the casualties of warfare. If we desire peace, we must consider the valence of our weapons and our attachment to war-making. If we desire peace, we must consider the face of those we might *other* through the design and implementation of technology.

CONCLUSION: GUNS TO GARDENS

From a theological (and philosophical) perspective, the object under ethical consideration is not the problem, per se. A piece of technology is typically inert without a human creator and operator. Technology, from an economic and political standpoint, is created because of its utilitarian function for a particular state or society. Those that wield the power, and have funding from the state, will most likely continue to exploit the public and private sectors stafor personal protection and gain. This is, after all, a part of our rebellion against nature (i.e., fallenness). What hope is there for us and for the environment that supports us?

Love of neighbor should be our guiding principle for all things we design and create. There is so much more to this theme than a mere platitude to *be nice*. We love our neighbor when we refuse to create weapons of mass destruction. We love our neighbor when we give them a face amid the hopelessness they may feel. We love our neighbor when in the middle of political conflict, we open our doors instead of barricading them. The highest probability for the moral use of current and future weapon systems, historically, comes from civilian concern. Indeed, but as womanist theology and ecofeminism have taught us, concern for the other is a concern for the individual, for we are intertwined on the molecular level.

It is no surprise that the US is apathetic to the implications of their technological demands. We praise our *freedom* and pay a high political cost for such and do not consider the implication for those outside of our time zone. Our commitment to big data, AI, and robotics comes at a cost to the world. China's surveillance state is no accident and most likely the future of the US. Currently, the concerns of surveillance by US citizens are low, and this puts many people in danger of exploitation and oppression. Since 9/11, congress has had little trouble convincing society that privacy is not our biggest concern, hence the passing of the Patriot Act in little over a month's time.[328] Juxtaposed with the growing global concern over digital identity or more specifically sovereign identity' framework (SSI), it is not hard to imagine the negative implications of technological creep and growing technocrats.

Think about the irony here; we are concerned about the use of AI in violent technologies in the context of foreign warfare but are completely apathetic to the NSA using our data to build systems of manipulation, domination, and totalitarianism. Just because a human is in the loop does not make the moral and ethical implications any less concerning. Problematic use of artificially intelligent algorithms occur in law enforcement, healthcare, and commerce. The human is the primary problem and not necessarily the solution. Human law codes and the concern about torts appear quite early in human history and there is good reason for that: humans are not self-regulating when it comes to morality and responsibility.[329] It did not take but a few hundred years, post the invention of writing, for law codes to emerge and remain basically the same for over 3,000 years.[330] What is unique and fascinating about the Judaic recapitulation of the typical ancient Near Eastern law-code is a prohibition of idol worship. The Hebrew word used in the second commandment *pesel* ("idol") is of particular importance here. The

---

[328] U.S. Congress, "Uniting and Strengthening America by Providing Appropriate Tools Required to Intercept and Obstruct Terrorism" (USA Patriot Act) Act of 2001, Public Law 107–56, https://www.congress.gov/107/plaws/publ56/PLAW-107publ56.pdf, (accessed Nov 21, 2022).

[329] There are some scholars, like James Q. Wilson who might disagree and state that because more citizens follow the law than break the law, this assumes that a moral sense is a universal and natural part of human sentiment per evolution. I find Wilson's work glib. Everyone breaks the law, but not to the same degree. Hence, why the law exists. It would be more accurate to say that most people are not murders or rapists, but to claim, as Wilson does in *The Moral Sense* that "most people are not criminals" is misleading. Wilson, *The Moral Sense* (New York: Free Press, 1993). Cf. Harvey A. Silverglate, *Three Felonies a Day: How the Feds Target the Innocent*, rev. ed. (New York: Encounter Books, 2011). Silverglate's premise is that vague laws lead to much blue-collar crime.

[330] See R. Westbrook, ed., *A History of Ancient Near Eastern Law*, vol. 1 (Atlanta, GA: SBL Press, 2017).

prohibition of image making in religious spaces is related to the posture of what the Bible often refers to as the "heart" (*leb*). It is an ANE expression of the innermost desires of the human. Love is demonstrated through the keeping of the commands (Deut 5:10; 10:12–13; Neh 1:5; Dan 9:4; John 14:15). Yes, in a literal sense of the reading of the Pentateuch, the prohibition of idol making is about proper worship (i.e., having the right *image*). However, the witness of Israel's history points to a more poetic reality, the idol lures the heart of the worshiper away from singular devotion. The story of the nation of Israel, according to both the Hebrew Bible and *Talmud*, is a tension and war with the idols of other gods. But the Israelites were not alone in critiquing the power of idols. Francis Bacon also wrestled with the idols of the mind: "the formation of ideas and axioms by true induction is no doubt the proper remedy to be applied for the keeping off a declaring away of idols."[331] Idols, according to Bacon, are a barrier to the scientific endeavor. The human mind is suspectable to false impressions, mistakes, and Bacon's god induction.

Karl Marx used the term "idol" in his discourse about money. He writes, "[m]oney degrades all the gods of mankind and converts them into commodities."[332] The creations of humankind are a tempting serpent to lure us away from the true life, alienating us through a fetish for the product. It is in many ways the return to paganism that Nietzsche called for, the turning from the ideals of Christian virtues of self-denial, poverty, meekness, and humility. In his *Twilight of the Idols* he writes, "The saint in whom God takes pleasure is the ideal castrate…Life is at an end whenever the kingdom of God begins." One can appreciate Nietzsche's skepticism here, and likewise connect the notion that reason is linked to desire. Regardless of confession or belief, idolatry is about relatedness (or identity). The Ten Commandments begin with a relationship, and thus not merely a set of rules. Why? Because who we are in relation to others shapes our identity. Richard Lints puts it this way, "[h]uman significance rests precisely in what or whom humans reflect and the relationship in which those reflections occur."[333]

331 Francis Bacon, *The Essays* (Harmondsworth: Penguin Books, 1985), 277.

332 Karl Marx, *Writings of the Young Marx on Philosophy and Society*, trans. L. D. Easton and K. H. Cuddat (New York: Doubleday, 1967), 246.

333 Richard Lints, *Identity and Idolatry: The Image of God and Its Inversion*, NSBT 36, (Downers Grove, IL: IVP Academic Press, 2015), 155.

The US attachment to the war machine is an identity problem. If we desire to be peacemakers, as we have seen with biological weapons,[334] we must limit the production and usage of violent tech. We not only have an addition to war in the West, but we also have a broader addiction to individual rights and freedoms, no matter the expense to our geographical or political neighbor. Acceptance of extreme forms of violence, like war, are but a symptom of a larger issue here in the United States of America. We have far more guns than we have people, and heaven only knows how many bullets accompany these weapons.[335] So what are we to do with the time we have left?

One of the reasons I have highlighted feminist and womanist theologies in some of the conclusions is to show that there is hope in the struggle ahead. Mainstream approaches to technology can be agnostic at times about what the consumer should do. Unlike the $11 billon dollar self-help industry which states the problems we face find their solution internally, and the male driven bias of the tech world which seeks domination over nature, I believe we should *think otherwise*.[336] As a pastor, I have worked with people who struggle with addictions, identity issues, and all of the classic existential questions in between. I know first-hand, and from clinical psychology, that telling an individual or community that the solution to their struggle is found within willpower or self-discipline is to set them up for failure. It takes outside influence to make change. The voice of experience in overcoming violence (in all of its forms) is found in historical analysis. May we listen to the voices of the past, who were denied a *face* (i.e., violated) so that we may learn from them and live in their wisdom. Our future with technology is not fated with the violent technocrats of science fiction. We have a choice. Do we want to use our research and funding to make weapons or to make gardens?

---

334 See Glen Stassen, "Just Peacemaking as Hermeneutical Key: The Need for International Cooperation in Preventing Terrorism," *Journal of the Society of Christian Ethics* 24, no. 2 (2004): 171–91.

335 Thomas Black, "Americans Have more Guns than Anywhere Else in the World and They Keep Buying More," *Bloomberg*, May 25, 2022, www.bloomberg.com/news/articles /2022-05-25/how-many-guns-in-the-us-buying-spree-bolsters-lead-as-most-armed-country, (accessed Nov 21, 2022).

336 I borrow this phrase from the work and logic of David Gunkel. See works mentioned above.

APPENDIX

Peshitta

And Cain said to Abel his brother, "Let us go to the valley." And it came to pass that, while they were in the field, Cain rose up against Abel his brother and killed him.

Targum Onkelos

And Cain said to his brother Abel…and it happened when they were in the field, and Cain rose up against Abel his brother and killed him.

Cario Geniza

And Cain said to his brother Abel, "Come, let the two of us go out to the open field." And it was when the two of them had gone out to the open field, [Cain] answered and said to Abel, "I have observed that the world was created in mercy, and it is guided in mercy.

Fragment Targum

And Cain said to Abel his brother, "Come, let the two of us go out to the open field." And it was when the two of them had gone out to the open field, Cain answered and said to Abel, "There is no judgment, and there is no judge, and there is no other world, and there is no gift of good reward for the righteous, and there is no punishment for the wicked. And the world was not created in mercy and it is not governed in mercy. For what reason was your offering accepted from you with favor, but from me it was not accepted with favor?" Abel answered and said to Cain, "There is judgment and there is a judge and there is another world, and there is a gift of good reward for the righteous, and punishment for the wicked. The world was created in mercy and it is governed in mercy. Because the fruits of [my] good works were better than yours, my offering was accepted from me with favor but from you it was not accepted with favor." And the two of them were arguing in the open □ eld and Cain rose up against Abel his brother and killed

Targum Neofiti

And Cain said to Abel his brother, "Come, let the two of us go out to the open field." And when the two of them went out into the open field, Cain answered and said to Abel, "I understand that the world was not created in mercy and is not governed according to the fruits of good works, and that there is partiality in judgment. For what reason was your offering accepted with favor, but my offering was not accepted from me with favor?" Abel answered and said to Cain, "I understand that the world was created in mercy and according to the fruits of good works it is governed. And it was because my deeds were better than yours that my offering was received from me with favor. Your offering was not accepted with favor

from you." Cain answered and said to Abel, "There is no judgment, and there is no judge, and there is no other world. There is no gift of good reward for the righteous and there is no punishment for the wicked." Abel answered and said to Cain, "There is judgment and there is a judge, and there is another world, and there is a gift of good reward for the righteous, and there is punishment for the wicked in the world to come." On account of this thing, the two of them were arguing in the open field, and Cain rose up against Abel his brother and killed him.

FIGURES & TABLES

ACKNOWLEDGEMENTS

Books are never finished products. They are much like children, we raise them until a certain point and then we must let go. This is one child I have worried about sending out into the world. I looked for help and insight from many authors and the generosity I received did not disappoint. I would like to express my deepest appreciation to Elke Schwarz for her early support of this project and the guidance she offered. I am grateful for Helen Paynter's early feedback on chapters 1 and 2. Thanks to my friend Kate Ott for her early reading and comments on the early draft of this project. I am grateful for Steven Umbrello and his faith in this project from the beginning.

All of this would not be possible without the support, love, and faith of my dearest friend and partner, Lucy. My dear, thank you for the years of love and hope that you have given to me. Thank you for nourishing my soul in the darkest of nights. I can face the dark because I know you walk beside me. Forgive me for the many hours of neglect during the writing of this book.

BIBLIOGRAPHY

Abelard, Peter, and Sandro Buzzetti. *Sententie magistri Petri Abelardi (Sententie Hermanni).* 1. ed. Pubblicazioni della Facoltà di lettere e filosofia dell'Università di Milano; Sezione a cura dell'Istituto di storia della filosofia, 101. 31. Firenze: La Nuova Italia, 1983.

"Activities of the Research and Development Center – Thailand, Advanced Research Projects Agency," n.d.

Adler, Mortimer. *The Conditions of Philosophy: Its Checkered Past, Its Present Disorder, and Its Future Promise.* New York: Dell Publishing Co, 1967.

Aizaid, Hugh, and Rosa Brooks. "Recommendations and Report of the Task Force on US Drone Policy." Stimson Center, April 2015. www.stimson.org/wp-content/files/file-attachments/recommendations_and_report_of_the_task_force_on_us_drone_policy_second_edition.pdf.

Anderson, Craig. *Roman Law for Scots Law Students.* Edinburgh: Edinburgh University Press, 2021.

Andreason, N. E. *The Old Testament Sabbath A Tradition: Historical Investigation.* SBL Dissertation Series 7. Missoula, MN: Scholars Press, 1972.

Antic, Radiša. "Can, Abel, Seth, and the Meaning of Human Life as Portrayed in the Books of Genesis and Ecclesiates." *Andres University Seminary Studies* 44 (2006): 203–11.

Aploinario, Tisha. "How Many People Play Call of Duty? (User & Growth Stats 2022." Fiction Horizon, Ocotber 2022. https://fictionhorizon.com/how-many-people-play-call-of-duty-user-growth-stats.

Arendt, Hannah. *On Violence.* New York: Harcourt, Brace & World, 1970.

Defense Advanced Research Projects Agency Website. "ARPA Is Born," n.d. www.darpa.mil/about-us/timeline/dod-establishes-arpa.

Asaro, Peter. "On Banning Autonomous Weapon Systems: Human Rights, Automation, and the Dehumanization of Lethal Decision-Making." *International Review of the Red Cross* 94, no. 886 (June 2012): 687–709. https://doi.org/10.1017/S1816383112000768.

Bacon, Francis. *The Essays.* Classics. Harmondsworth: Penguin Books, 1985.

Bahrani, Zainab. *Rituals of War: The Body and Violence in Mesopotamia.* New York: Cambridge: Zone Books; Distributed by The MIT Press, 2008.

Baker-Fletcher, Karen. *Dancing with God: The Trinity from a Womanist Perspective.* St. Louis, Mo: Chalice Press, 2006.

Barr, David L., ed. *Reading the Book of Revelation: A Resource for Students.* Resources for Biblical Study, no. 44. Atlanta: Society of Biblical Literature, 2003.

Barron, Andrew. "Depiction of War in Games: Can You Do Better?" Presented at the Global Gaming Conference, 2018. https://mail.google.com/mail/u/1/#inbox/FFNDWMkjVFHsRxfVrkXRjGQZtTKvBXvh?projector=1.

Barth, Karl, Geoffrey W. Bromiley, and Karl Barth. *The Doctrine of God. p. 2: Vol. 2. The Doctrine of God.* Vol. 2. London: Clark, 2004.

Baumeister, Roy F. *Evil: Inside Human Violence and Cruelty.* 1. ed. New York: Holt Paperbacks, 2001.

Beale, G. K. *A New Testament Biblical Theology: The Unfolding of the Old Testament in the New.* Grand Rapids: Baker Academic, 2011.

Beck, Aaron T. *Prisoners of Hate: The Cognitive Basis of Anger, Hostility, and Violence.* New York: Harper, 2000.

Belfiore, Michael P. *The Department of Mad Scientists: How DARPA Is Remaking Our World, from the Internet to Artificial Limbs.* 1st ed. [Washington, D.C.]: New York: Smithsonian Books; Harper, 2009.

Benjamin, Ruha. *Race after Technology: Abolitionist Tools for the New Jim Code.* Medford, MA: Polity, 2019.

Benjamin, Walter, Peter D. Fenves, and Julia Ng. *Toward the Critique of Violence: A Critical Edition.* Stanford (Calif.): Stanford university press, 2021.

Benso, Silvia. *The Face of Things: A Different Side of Ethics*. SUNY Series in Contemporary Continental Philosophy. Albany, N.Y: State University of New York Press, 2000.

Berkouwer, G. C. *Faith and Sanctification*. Grand Rapids, Mich.: Eerdmans, 1952.

Bettenson, Henry, and Chris Maunder, eds. *Documents of the Christian Church*. 4th ed. / edited by Chris Maunder. New York: Oxford University Press, 2011.

Biemann, Asher. "Humanizing the It: Martin Buber on Technology and the Ethics of Things." *Religions* 13, no. 137 (2022): 1–25.

Black, Thomas. "Americans Have More Guns than Anywhere Else in the World and They Keep Buying More." *Bloomberg*, May 25, 2022. https://www.bloomberg.com/news/articles/2022-05-25/how-many-guns-in-the-us-buying-spree-bolsters-lead-as-most-armed-country.

Bonhoeffer, Dietrich. *Dietrich Bonhoeffer Works. Volume 3: Creation and Fall: A Theological Exposition of Genesis 1-3 / Dietrich Bonhoeffer; Translated from the German Edition Edited by Martin Rüter and Ilse Tödt; English Edition Edited by John W. de Gruch ; Translated by Douglas Stephen Bax*. Edited by Martin Rüter, Ilse Tödt, and John W. De Gruchy. Translated by Douglas S. Bax. First Fortress Press paperback edition. Minneapolis: Fortress Press, 2004.

Bonvillian, W. B. "Evolution of U.S. Government Innovation Organization: From the Pipeline Model to the Connected Model, to the Problem of 'Political Design." Presented at the GRIPS Innovation, Science, and Technology Seminar, Tokyo, April 2013.

Bostrom, Nick. "A History of Transhumanist Thought." *Journal of Evolution and Technology* 14, no. 1 (2005): 1.

———. *Superintelligence: Paths, Dangers, Strategies*. Oxford, United Kingdom; New York, NY: Oxford University Press, 2016.

Botting, Eileen Hunt. *Artificial Life after Frankenstein. Artificial Life after Frankenstein*. 1st edition. Philadelphia: University of Pennsylvania Press, 2021.

Breasted, James, and Thomas G. Allen. "The Historical Record of Rameses III: The Texts in Medinet Habu." In *Studies in Ancient Oriental Civilization*, Vol. I–II. The Oriental Institute of the University of Chicago. Chicago, IL: University of Chicago Press, 1956.

Brue, Ethan J., Derek Schuurman, and Steven H. VanderLeest. *A Christian Field Guide to Technology for Engineers and Designers*. Downers

Grove, Illinois: IVP Academic, An Imprint of InterVarsity Press, 2022.

Brunner, H. G., M. Nelen, X. O. Breakfield, and H. B. A. van Oost. "Abnormal Behavior Associated with a Point Mutation in the Structural Gene for Monoamine Oxidase A." *Science* 262 (1993): 578–80.

Buber, Martin. *Schriften Über Das Dialogische Prinzip*. Auflage. Buber, Martin. Werkausgabe 4. Gütersloh: Gütersloher Verlagshaus, 2019.

Budge, E. A. Wallis. *Annals of the Kings of Assyria: The Cuneiform Texts with Translations and Transliterations from the Original Documents*. London: Kegan Paul, 2007.

Bufacchi, Vittorio, ed. *Violence: A Philosophical Anthology*. Basingstoke: Palgrave Macmillan, 2009.

Bush, George. "State of the Union Address." *Washington Post*, January 28, 2003. www.washingtonpost.com/wp-srv/onpolitics/transcripts/bushtext_012803.html.

Butler, Judith. *Frames of War: When Is Life Grievable?* Radical Thinkers. London New York: Verso, 2016.

Butler, Judith Pamela. *The Force of Nonviolence: An Ethico-Political Bind.* London New York: Verso, 2020.

Calvin, Jean, Ford Lewis Battles, and John T. McNeill. *Calvin: Institutes of the Christian Religion*. Paperback edition. Louisville, Kentucky: Westminster John Knox Press, 2011.

Camus, Albert, and Justin O'Brien. *Resistance, Rebellion, and Death*. 1st Vintage International ed. New York: Vintage International, 1995.

Carpenter, Julie. *Culture and Human-Robot Interaction in Militarized Spaces: A War Story*. Emerging Technologies, Ethics and International Affairs. Burlington, VT: Ashgate Publishing Company, 2016.

Carroll R., M. Daniel, and J. Blair Wilgus, eds. *Wrestling with the Violence of God: Soundings in the Old Testament*. Bulletin for Biblical Research Supplements. Winona Lake, Indiana: Eisenbrauns, 2015.

Cavanaugh, William. "Does Religion Cause Violence?" *Harvard Divinity Bulletin* 35, no. 2/3 (2007): 1–14.

Chalmers, David John. *The Conscious Mind: In Search of a Fundamental Theory*. 1. issued as an Oxford University Press paperback. Philosophy of Mind Series. New York: Oxford University Press, 1997.

Chappelle, Wayne. "An Analysis of Post-Traumatic Stress Symptoms in Unites States Air Force Drone Operators." *Journal of Anxiety Disorders* 28, no. 5 (2014): 480–87.

Chermol, B. H. "Wounds Without Scars: Treatment of Battle Fatigue in the U.S. Armed Forces in the Second World War." *Military Affairs: The Journal of Military History, Including Theory and Technology* 49 (1985): 9–12.

WHO. "Child Mortality and Causes of Death," n.d. www.who.int/data/gho/data/themes/topics/topic-details/GHO/child-mortality-and-causes-of-death.

Chomsky, Noam. *Language and Problems of Knowledge: The Managua Lectures*. Current Studies in Linguistics Series 16. Cambridge, Mass: MIT Press, 1988.

Churchland, Paul. "Eliminative Materialism and the Propositional Attitude." *Journal of Philosophy* 78, no. 2 (1981).

Clausewitz, Carl von. *On War*. Berlin: The Floating Press, 2010.

Coeckelbergh, Mark. *Growing Moral Relations: Critique of Moral Status Ascription*. New York: Palgrave Macmillan, 2012.

———. *Robot Ethics*. The MIT Press Essential Knowledge Series. Cambridge, Massachusetts: The MIT Press, 2022.

Coker, Christopher. *Waging War without Warriors? The Changing Culture of Military Conflict*. IISS Studies in International Security. Boulder, Colo: Lynne Rienner Publishers, 2002.

Collins, John. "The Zeal of Phinehas: The Bible and the Legitimation of Violence." *Society of Biblical Literature* 122, no. 1 (2003): 3–21.

Collins, Randall. *Violence: A Micro-Sociological Theory*. Princeton, NJ: Princeton Univ. Press, 2008.

Cone, James H. *God of the Oppressed*. Rev. ed. Maryknoll, N.Y: Orbis Books, 1997.

Cone, James H., and Gayraud S. Wilmore, eds. *Black Theology: A Documentary History*. 2nd ed., Rev. Maryknoll, N.Y: Orbis Books, 1993.

Cottrill, Amy C. *Uncovering Violence: Reading Biblical Narratives as an Ethical Project*. First edition. Louisville, Kentucky: Westminster John Knox Press, 2021.

Coudenhove-Kalergi, Richard. *Eine Idee Erobert Europa: Meine Lebenserinnerungen*. Vienna: Kurt Desch, 1958.

Cowles, C. S., ed. *Show Them No Mercy: Four Views on God and Canaanite Genocide*. Counterpoints. Grand Rapids, Mich: Zondervan, 2003.

Cramer, David C., and Myles Werntz. *A Field Guide to Christian Nonviolence: Key Thinkers, Activists, and Movements for the Gospel of Peace*.

Grand Rapids, Michigan: Baker Academic, a division of Baker Publishing Group, 2022.

———. *A Field Guide to Christian Nonviolence: Key Thinkers, Activists, and Movements for the Gospel of Peace.* Grand Rapids, Michigan: Baker Academic, a division of Baker Publishing Group, 2022.

Crippen, Linda. "Military Working Dogs: Guardians of the Night." Army.mil, 2017. www.army.mil/article/56965/military_working_dogs_guardians_of_the_night.

Crootof, Rebecca. "Killer Robots Are Here: Legal and Policy Implications." *Cardozo Law Review* 36, no. 5 (2014): 1837.

———. "War Torts." *New York Law Review* 97 (2022).

Crootof, Rebecca, Margot E. Kaminski, and Nicholson Price. "Humans in the Loop." *Vanderbilt Law Review*, 2023.

Curse, Sarah. "Military Working Dogs: Classification and Treatment in the U.S. Armed Forces." *Animal Law* 21, no. 2 (2015): 249–89.

ICRC. "Customary IHL - Rule 70. Weapons of a Nature to Cause Superfluous Injury or Unnecessary Suffering," n.d. www.icrc.org/customary-ihl/eng/docs/v1_rul_rule70.

Darling, Kate. *The New Breed: What Our History with Animals Reveals about Our Future with Robots.* First edition. New York: Henry Holt and Company, 2021.

Darpa.Mil. "Darpa Robotics Challenge," n.d. www.darpa.mil/program/darpa-robotics-challenge.

Davis, Erik. *TechGnosis: Myth, Magic, & Mysticism in the Age of Information.* Berkeley, California: North Atlantic Books, 2015.

DeGrazia, David. *Taking Animals Seriously: Mental Life and Moral Status.* Cambridge; New York: Cambridge University Press, 1996.

Demarest, Bruce A. *The Cross and Salvation: The Doctrine of God.* Foundations of Evangelical Theology. Wheaton, Ill: Crossway Books, 2006.

Der Derian, James. *Virtuous War: Mapping the Military-Industrial-Media-Entertainment Network.* 2nd ed. New York: Routledge, 2009.

Diamond, James. "The Deuteronomic 'Pretty Woman' Law: Prefiguring Feminism and Freud in Nahmanides." *Jewish Social Studies* 14, no. 2 (2008): 61–85.

"Directive 3000.09." Department of Defense, n.d. www.esd.whs.mil/portals/54/documents/dd/issuances/dodd/300009p.pdf.

Dobos, Ned. *Ethics, Security, and The War Machine.* Oxford University Press, 2022.

Dowe, Phil. *Physical Causation.* Cambridge Studies in Probability, Induction, and Decision Theory. Cambridge; New York: Cambridge University Press, 2000.

Dunfee, Susan N. "The Sin of Hiding." *Soundings* 65, no. 3 (1982): 316–27.

Ehrman, Albert. "What Did Cain Say to Abel?" *Jewish Quarterly Review*, no. 53 (1962): 164–67.

Eibl-Eibesfeldt, Irenaeus. *Love and Hate: The Natural History of Behavior Patterns.* 1st Aldine pbk. ed. Foundations of Human Behavior. New York: Aldine de Gruyter, 1996.

Eichrodt, Walther, Hans Wilhelm Hertzberg, Martin Noth, James Luther Mays, F. Alberto Soggin, William McKane, Otto Kaiser, David L. Petersen, Norman C. Habel, and Robert P. Carroll. *The Old Testament Library.* Philadelphia: Westminster Press, 1961.

Eisenhower, Dwight. "Address to the American Society of Newspaper Editors." The American Presidency Project, April 16, 1953. https://www.presidency.ucsb.edu/documents/address-the-chance-for-peace-delivered-before-the-american-society-newspaper-editors,

———. "President Dwight D. Eisenhower's Farewell Address." National Archives Catalog, n.d. https://catalog.archives.gov/id/594599.

Ellul, Jacques. *Violence: Reflections from A Christian Perspective.* Eugene, Or: Wipf & Stock, 2007.

Ellul, Jacques, and John Wilkinson. *The Technological Society: A Penetrating Analysis of Our Technical Civilization and of the Effect of an Increasingly Standardized Culture on the Future of Man.* [Nachdruck der Ausgabe] New York, Knopf. A Vintage Book. New York, NY: Vintage books, 2011.

Elman, Peral. "Deuteronomy 21:10–14: The Beautiful Captive Woman." *Women in Judaism* 1 (1997): 1–13.

Emery, John. "Probabilities Towards Death: Bugsplat, Algorithmic Assassinations, and Ethical Due Care." *Critical Military Studies*, n.d. https://doi.org/10.1080/23337486.2020.1809251.

Erickson, Millard J. *Christian Theology.* 3rd ed. Grand Rapids, Mich: Baker Academic, 2013.

———. *Christian Theology*. 3rd ed. Grand Rapids, Mich: Baker Academic, 2013.

Ernst, Katie. "Terror of Hope: The Integration of Attachment Theory and Womanist Theology." *Pastoral Psychology* 66 (2017): 201–12.

Fanon, Frantz, Richard Philcox, Homi K. Bhabha, Jean-Paul Sartre, and Cornel West. *The Wretched of the Earth*. 60th anniversary edition. New York: Grove Press, 2021.

Farrar, Frederic. *The History of Interpretation*. London: Macmillan and Co., 1886.

Feyerabend, Paul. *Philosophy of Nature*. Edited by Helmut Heit and Eric Oberheim. Translated by Dorothea Lotter. Paperback edition. Cambridge, UK: Polity Press, 2018.

Finlay, Charles. "Hannah Arendt's Critique of Violence." *Thesis Eleven* 97 (2009): 26–45.

Flusser, Vilém. *The Shape of Things: A Philosophy of Design*. London: Reaktion, 1999.

Fortier, David. "The Adoption of the Black Rifle: Project AGILE: Colt's AR-15 Went from Utter Failure to Adoption in Just Three Years,"." *Firearm News* 74, no. 21 (2020): 26.

Foucault, Michel, and Michel Foucault. *Discipline and Punish: The Birth of the Prison*. 2. Vintage Books ed. New York: Vintage Books, 1995.

Frantzman, Seth. "As Drones Proliferate, Authoritarian Regimes Profit." *Newlines Institute*, June 23, 2021.

Frazer, Elizabeth, and Kimberly Hutchings. *Violence and Political Theory*. And Political Theory. Cambridge, UK; Medford, MA: Polity, 2020.

French, Shannon E. *The Code of the Warrior: Exploring Warrior Values Past and Present*. Lanham, Md: Rowman & Littlefield Publishers, 2003.

Gale, Zona. *Miss Lulu Bett*. Middlesex: Echo Library, 2007.

Garsiel, Mose. *Biblical Names: A Literary Study of Midrashic Derivation and Puns*. Ramat Gan: Bar-Ilan University Press, 1991.

Gattini, Andrea. "To What Extent Are State Immunity and Non-Justiciability Major Hurdles to Individuals' Claims for War Damages?" *Journal of International Criminal Justice* 1, no. 2 (n.d.): 348–51.

Girard, René, Yvonne Freccero, and René Girard. *Deceit, Desire, and the Novel: Self and Other in Literary Structure*. Nachdr. Baltimore: Johns Hopkins University Press, 1998.

Girard, René. *The Scapegoat.* Johns' Hopkins paperbacks ed. Baltimore: Johns Hopkins University Press, 1989.

———. *Things Hidden since the Foundation of the World.* Repr. Stanford, Calif: Stanford University Press, 2000.

———. *Violence and the Sacred.* John Hopkins paperbacks ed., [Nachdr.]. Baltimore: Hopkins Univ. Press, 1979.

Godfrey-Smith, Peter. *Other Minds: The Octopus, the Sea, and the Deep Origins of Consciousness.* First edition. New York: Farrar, Straus and Giroux, 2016.

Goldingay, John. *Old Testament Theology.* Downers Grove, Illinois: IVP Academic, 2016.

Goldstein, Valerie. "The Human Situation: A Feminine View." *Journal of Religion* 40 (1960): 100–112.

Gould, Stephen Jay. *Full House: The Spread of Excellence from Plato to Darwin.* First paperback edition. New York: Three Rivers Press, 1997.

Grant, Jacquelyn. *White Women's Christ and Black Women's Jesus: Feminist Christology and Womanist Response.* American Academy of Religion Academy Series, no. 64. Atlanta, Ga: Scholars Press, 1989.

Grossman, Dave. *On Killing: The Psychological Cost of Learning to Kill in War and Society.* Rev. ed. New York: Little, Brown and Co, 2009.

Gubrud, Mark. "Stopping Killer Robots." *Bulletin of the Atomic Scientists* 70, no. 1 (2014): 32.

Gunkel, David. *The Machine Question: Critical Perspecitves on AI, Robots, and Ethics.* Cambridge, Massachusetts: The MIT Press, 2012.

———. *Robot Rights.* Cambridge, Massachusetts: The MIT Press, 2018.

———. *Person, Thing, Robot: An Ontology for the 21st Century.* Cambridge, Massachusetts: The MIT Press 2023.

———. "The Rights of (Killer) Robots." In in Steven C. Roach and Amy E. Eckert (eds.) Moral Responsibility in 21st Century Warfare: Just War Theory and the Ethical Challenges of Autonomous Weapon Systems. SUNY, 2020.

Gultong, Johan. "Violence, Peace, and Public Research." *Journal of Peace Research* 6, no. 3 (1969): 168.

Gusterson, Hugh. *Drone: Remote Control Warfare.* First MIT paperback edition. Cambridge, Massachusetts London: The MIT Press, 2017.

Hallo, William. "New Moons and Sabbaths: A Case-Study in the Contrastive Approach." *Bible and Spade* 9, no. 3/4 (1977): 1–18.

Hamer, Fannie Lou, Maegan Parker Brooks, and Davis W. Houck. *The Speeches of Fannie Lou Hamer: To Tell It like It Is*. Jackson: University Press of Mississippi, 2011.

Hamerton-Kelly, Robert. *Sacred Violence: Paul's Hermeneutic of the Cross*. Minneapolis: Fortress Press, 1992.

Hamilton, Richard Paul, and Margaret Sönser Breen, eds. *This Thing of Darkness: Perspectives on Evil and Human Wickedness*. At the Interface/Probing the Boundaries, v. 7. Amsterdam: Rodopi, 2004.

Harnack, Adolf von. *Militia Christi: The Christian Religion and the Military in the First Three Centuries*. Philadelphia: Fortress Press, 1981.

Heggen, Carolyn Holderread. *Sexual Abuse in Christian Homes and Churches*. Eugene, Oregon: Wipf & Stock, 2006.

Heidegger, Martin. *The Question Concerning Technology and Other Essays*. Translated by William Lovitt. Harper Perennial Modern Thought. New York; London Toronto: HarperCollins Publishers, 2013.

Heidegger, Martin, and Gregory Fried. *Introduction to Metaphysics*. Second Edition. New Haven: Yale University Press, 2014.

Heilbrun, James, and Charles M. Gray. *The Economics of Art and Culture*. 2nd ed. New York: Cambridge University Press, 2001.

Heyns, Christof. "Armed Drones and the Right to Life." Special Report to General Assembly. UN, 2013. www.oxfordmartin.ox.ac.uk/publications/armed-drones-and-the-right-to-life/.

Hoerster, Norbert. *Neugeborene und das Recht auf Leben*.Aufl. Suhrkamp-Taschenbuch Wissenschaft 1215. Frankfurt am Main: Suhrkamp, 1995.

Holsaert, Faith S., ed. *Hands on the Freedom Plow: Personal Accounts by Women in SNCC*. Urbana: University of Illinois Press, 2010.

Hornus, J. M. *Évangile et Labarum*. Geneva: Labor & Fides, 1960.

Horsley, Richard A. *Jesus and the Spiral of Violence: Popular Jewish Resistance in Roman Palestine*. 1st Fortress Press ed. Minneapolis: Fortress Press, 1993.

Huntemann, Nina, and Matthew Thomas Payne, eds. *Joystick Soldiers: The Politics of Play in Military Video Games*. New York: Routledge, 2010.

Hylen, Susan E. "Metaphor Matters: Violence and Ethics in Revelation." *Catholic Biblical Quarterly* 73, no. 4 (2011): 777–96.

Jackelen, Antje. "The Image of God as Techno Sapiens." *Zygon* 37, no. 2 (2002): 296.

Jacobson, Howard. "Genesis 4:8." *Vetus Testamentum*, no. 55 (2005): 564–65.

Janzen, Mark. "The Iconography of Humiliation: The Depiction and Treatment of Bound Foreigners in New Kingdom Egypt." PhD Dissertation, University of Memphis, 2013.

Johnson, Keith L., ed. *T&T Clark Companion to the Doctrine of Sin*. New York: Bloomsbury T&T Clark, an imprint of Bloomsbury Publishing Place, 2016.

JR, King, Martin Luther. *Letters From Birmingham Jail*. Place of publication not identified: Penguin, Books, 2018.

Katz, Yarden. *Artificial Whiteness: Politics and Ideology in Artificial Intelligence*. New York: Columbia University Press, 2020.

Kieran, David, ed. *The War of My Generation: Youth Culture and the War on Terror*. London: Rutgers University Press, 2015.

Kierkegaard, Søren. *The Concept of Anxiety: A Simple Psychologically Oriented Deliberation in View of the Dogmatic Problem of Hereditary Sin*. Edited by Alastair Hannay. Liveright paperback. New York: Liveright Publishing Corporation, 2015.

Kim, Tae Wan Kim. "Should Robots Have Rights or Rites? A Confucian Cross-Cultural Exploration to Robot Ethics," 2020, 1–9.

King, Ynestra. "Engendering a Peaceful Planet: Ecology, Economy, and Ecofeminism in Contemporary Context." *Rethinking Women's Peace Studies* 23, no. 3/4 (1995): 15–21.

Kirby, Vicki. *Telling Flesh: The Substance of the Corporeal*. New York: Routledge, 1997.

Kirk, Alan. "Love Your Enemies,' The Golden Rule, and Ancient Reciprocity (Luke 6:27–35)." *Journal of Biblical Literature* 122, no. 4 (n.d.): 667–86.

Kirk-Duggan, Cheryl A. *Violence and Theology*. Horizons in Theology. Nashville: Abingdon Press, 2006.

Klassen, William. "Vengeance in the Apocalypse of John." *Catholic Biblical Quarterly* 28 (1966): 300–311.

Kline, Meredith. "Lex Talionis and the Human Fetus." *Journal of Evangelical Theological Studies* 20 (1977): 273–304.

Kripke, Saul A. *Naming and Necessity*. Oxford, UK; Cambridge, USA: Blackwell Publishers, 1998.

Krishnan, Armin. *Killer Robots: Legality and Ethicality of Autonomous Weapons*. Farnham, England; Burlington, VT: Ashgate, 2009.

Kurzweil, Ray. *The 10% Solution for a Healthy Life: How to Reduce Fat in Your Diet and Eliminate Virtually All Risk of Heart Disease and Cancer*. Paper. New York: Crown Publishers, 1993.

———. *The Age of Spiritual Machines: When Computers Exceed Human Intelligence*. New York, NY: Penguin Books, 2000.

———. *The Singularity Is near: When Humans Transcend Biology*. New York, NY: Penguin Books, 2006.

Kurzweil, Ray, and Terry Grossman. *Fantastic Voyage: Live Long Enough to Live Forever*. 1. Plume print. New York, NY: Plume, 2005.

———. *Transcend: Nine Steps to Living Well Forever*. New York: Rodale, 2011.

Lakoff, George, and Mark Johnson. *Metaphors We Live By*. Chicago: University of Chicago Press, 2003.

LaPorte, Joseph. *Rigid Designation and Theoretical Identities*. Oxford: Oxford University Press, 2013.

Latour, Bruno. *We Have Never Been Modern*. Cambridge, Mass: Harvard University Press, 1993.

Lawrence, Bruce B., and Aisha Karim, eds. *On Violence: A Reader*. Durham [NC]: Duke University Press, 2007.

Lehman-Wilzig, Sam N., ed. *Comparative Jewish Politics*. Bar-Ilan Departmental Research/Dept. of Political Studies. Ramat-Gan, Israel: Bar-Ilan University Press, 1981.

Leibniz, Gottfried Wilhelm, George R. Montgomery, and Gottfried Wilhelm Leibniz. *Discourse on Metaphysics and The Monadology*. Dover Philosophical Classics. Mineola, N.Y: Dover Publ, 2005.

Levinas, Emmanuel. *Difficult Freedom: Essays on Judaism*. Baltimore: Johns Hopkins University Press, 2012.

———. *Otherwise than Being, or, Beyond Essence*. Pittsburgh, Pa: Duquesne University Press, 1998.

———. *Totality and Infinity: An Essay on Exteriority*. 23rd printing. Pittsburgh, Pa: Duquesne Univ. Press [u.a.], 2011.

———, Adriaan Theodoor Peperzak, Simon Critchley, and Robert Bernasconi. *Emmanuel Levinas: Basic Philosophical Writings*. Studies in Continental Thought. Bloomington: Indiana University Press, 1996.

Lewis, David K. “An Argument for the Identity Theory.” *The Journal of Philosophy* 63, no. 1 (1966): 17–25.

Lints, Richard. *Identity and Idolatry: The Image of God and Its Inversion*. New Studies in Biblical Theology 36. Downers Grove, Illinois: InterVarsity Press, 2015.

Lohor, Joel. "Righteous Abel, Wicked Cain: Genesis 4:1–16 in the Masoretic Text, The Septuagint, and the New Testament." *Catholic Biblical Quarterly* 71 (2009): 494.

Luther, Martin, Walther I. Brandt, and Helmut T. Lehmann. *Luther's Works*. Philadelphia (Pa.): Fortress Press, 1962.

MacAskill, William. "The Case for Longtermism." *New York Times*, August 7, 2002. www.nytimes.com/2022/08/05/opinion/the-case-for-longtermism.html.

Marion, Jean-Luc. *Being given: Toward a Phenomenology of Givenness*. Cultural Memory in the Present. Stanford, Calif: Stanford University Press, 2002.

———. *Prolegomena to Charity*. First Edition. Perspectives in Continental Philosophy 24. New York: Fordham University Press, 2002.

Markey, Patrick M., and Christopher J. Ferguson. *Moral Combat: Why the War on Violent Video Games Is Wrong*. Dallas, TX: BenBella Books, 2017.

Marra, William, and Sonia K. McNeil. "Understanding 'The Loop': Regulating the Next Generation of War Machines." *Havard Journal of Law and Public Policy* 36 (2013): 1139–43.

Martínez, Florention. "Eve's Children in Foe Targumim." In *Eve's Children: The Biblical Stories Retold and Interpreted in Jewish and Christian Traditions*, 5:28–29. Kernes in Biblical Narrative. Leiden: Brill, 2003.

Marx, Karl, Loyd David Easton, and Kurt H. Guddat. *Writings of the Young Marx on Philosophy and Society*. Indianapolis, Ind: Hackett Pub. Co, 1997.

Matthews, Victor Harold, and Don C. Benjamin. *Old Testament Parallels: Laws and Stories from the Ancient Near East*. Fully rev. and Expanded 3rd ed. New York: Paulist Press, 2006.

McCall, Thomas H. *Against God and Nature: The Doctrine of Sin*. Foundations of Evangelical Theology Series. Wheaton, Ill: Crossway, 2019.

McEntire, M. "Being Seen and Not Hear: The Interpretation of Genesis 4:8." In *Of Scribes and Sages*, 9:4–13. Studies in Scripture in Early Judaism and Christianity. New York: T&T Clark, 2004.

McLuhan, Marshall. *The Medium and the Light: Reflections on Religion*. Edited by Eric McLuhan. Eugene, Or: Wipf & Stock, 2010.

———. *Understanding Media: The Extensions of Man*. 1st MIT Press ed. Cambridge, Mass: MIT Press, 1994.

Mendelsohn, Issac. *Slavery in the Ancient Near East: A Comparative Study on Slavery in Babylonia, Assyria, Syria, and Palestine from the Middle of the Third Millennium to the End of the First Millennium*. New York: Oxford University Press, 1949.

Mendenhall, George E., H. B. Huffmon, F. A. Spina, and Alberto Ravinell Whitney Green, eds. *The Quest for the Kingdom of God: Studies in Honor of George E. Mendenhall.* Winona Lake, Ind: Eisenbrauns, 1983.

Merrill, Eugene H. *Everlasting Dominion: A Theology of the Old Testament.* Nashville, Tenn.: Broadman & Holman, 2006.

Miller, Arthur G., ed. *The Social Psychology of Good and Evil.* Second edition. New York: The Guilford Press, 2016.

Minsky, Marvin. *The Emotion Machine: Commonsense Thinking, Artificial Intelligence, and the Future of the Human Mind.* 1. Simon & Schuster trade paperback ed. New York: Simon & Schuster, 2007.

Minsky, Marvin Lee. *The Society of Mind.* New-York: Simon and Schuster, 1988.

Moltmann, Jürgen. *Theology of Hope: On the Ground and the Implications of a Christian Eschatology*. 1st Fortress Press ed. Minneapolis: Fortress Press, 1993.

Moravec, Hans P. *Mind Children: The Future of Robot and Human Intelligence.* Cambridge, Mass: Harvard University Press, 1988.

———. *Robot: Mere Machine to Transcendent Mind.* First issued as an Oxford University Press paperback. Oxford New York Paris: Oxford University Press, 2000.

Murray, Michael J. *Nature Red in Tooth and Claw: Theism and the Problem of Animal Suffering.* Oxford; New York: Oxford University Press, 2008.

———. *Nature Red in Tooth and Claw: Theism and the Problem of Animal Suffering.* Oxford: Oxford University Press, 2011.

Neibuhr, Reinhold. *Christianity and Power Politics*. Hamden, CT: Archon Books, 1969.

Niebuhr, Reinhold. *The Nature and Destiny of Man: A Christian Interpretation.* 1st ed. Library of Theological Ethics. Louisville, Ky: Westminster John Knox Press, 1996.

Noble, David F. *The Religion of Technology: The Divinity of Man and the Spirit of Invention*. New edition, with A new preface. Penguin Book Religion Science. New York: Penguin Books, 1999.

Noble, Safiya Umoja. *Algorithms of Oppression: How Search Engines Reinforce Racism*. New York: New York University Press, 2018.

Noort, Ed. "Genesis 4:1–16: From Paradise to Reality: The Myth of Brotherhood." *Eve's Children*, n.d.

Nowrot, Karsten. "Animals in War: At the Vanishings Point of International Humanitarian Law." International Review of the Red Cross, n.d. https://international-review.icrc.org/articles/animals-in-war-at-the-vanishing-point-of-international-humanitarian-law-919#footnote13_3kfh9s2.

O'Conner, Erin. "Fraction of Men": Engendering Amputation in Victorian Culture." *Comparative Studies in Society and History* 39, no. 4 (1997): 742–67.

O'Neil, Cathy. *Weapons of Math Destruction: How Big Data Increases Inequality and Threatens Democracy*. First edition. New York: Crown, 2016.

Orlinsky, Harry. *Notes on the Translation of the Torah*. Philadelphia: Jewish Publication Society, 1969.

Osborne, Grant. *Revelation*. Baker Exegetical Commentary on the New Testament. Grand Rapids, Mich: Baker Academic, 2002.

Ovitt, George. *The Restoration of Perfection: Labor and Technology in Medieval Culture*. New Brunswick: Rutgers University Press, 1987.

Pelman, Alik. "Metaphysics of Pain; Semantics of 'Pain'." *Ratio* 28, no. 3 (2015): 148–53.

Raytheon Missiles & Defense. "Phalanx Weapon System," n.d. www.raytheonmissilesanddefense.com/what-we-do/naval-warfare/ship-self-defense-weapons/phalanx-close-in-weapon-system.

Philips, Dave, and Tim Arango. "Who Signs Up to Fight? Makeup of U.S. Recruits Shows Glaring Disparity." *New York Times*, January 10, 2020. www.nytimes.com/2020/01/10/us/military-enlistment.html.

Physics and Cosmology. 1. Vatican City State: Vatican Observatory Publications, 2007.

Plaskow, Judith. *Sex, Sin, and Grace: Women's Experience and the Theologies of Reinhold Niebuhr and Paul Tillich*. Washington: University Press of America, 1980.

Ponte, Susanne de, and Hubert Sowa. *Ereignis Und Wahrnehmung: Eine Interdisziplinäre Untersuchung Zu Den Events Der Künstlergruppe "Gang Art."* Saecula Spiritalia, Bd. 31. Baden-Baden: Koerner, 1996.

Gov Info. "Public Law 85-567–July 28, 1958." Goverment. Accessed November 11, 2022. www.govinfo.gov/content/pkg/STATUTE-72/pdf/STATUTE-72-Pg426-2.pdf.

Radner, Karen. *Die Macht Des Namens: Altorientalische Strategien Zur Selbsterhaltung.* Santag: Arbeiten Und Untersuchungen Zur Keilschriftkunde, Bd. 8. Wiesbaden: Harrassowitz, 2005.

Raine, Adrian. *The Anatomy of Violence: The Biological Roots of Crime.* First Vintage books edition. New York: Vintage Books, 2014.

Ravel, Juliane, and Madalena Rosa. "IHL in Action: Seven Patterns of Respect." *Humanitarian Law*, December 2, 2021. https://blogs.icrc.org/law-and-policy/2020/11/19/ihl-in-action-seven-patterns-respect.

The Economist. "Recreating ARPA, The Most Successful Research Agency in History," January 1, 2020. www.economist.com/britain/2020/01/23/recreating-arpa-the-most-successful-research-agency-in-history.

Renic, Neil C. *Asymmetric Killing: Risk Avoidance, Just War, and the Warrior Ethos.* First edition. New York, NY: Oxford University Press, 2020.

Renic, Neil, and Sebastian Kaempf. "Modern Lawfare: Exploring the Relationship between Military First-Person Shooter Video Games and the 'War Is Hell' Myth." *Global Studies Quarterly* 2 (2022): 1–11.

Resident Evil 7 Gore Compilation–Resident Evil 7 Goriest Moments–Resident Evil 7 Gory Death. YouTube, 2017. www.youtube.com/watch?v=PFD-Ym7_kQ0.

Resnick, David. "A Case Study in Jewish Moral Education: (Non)-Rape of the Beautiful Captive." *Journal of Moral Education* 33, no. 3 (2004): 307–19.

Rhodes, Richard, ed. *The Making of the Atomic Bomb.* 25th anniversary ed. New York, N.Y.: Simon & Schuster Paperbacks, 2012.

Riches, David, ed. *The Anthropology of Violence.* New York, NY, USA: Blackwell, 1986.

Roberts, J. Deotis. *Liberation and Reconciliation: A Black Theology.* 2nd ed. Louisville, Ky: Westminster John Knox Press, 2005.

Robot Rules. New York, NY: Springer Berlin Heidelberg, 2018.

Rosenberg, Alexander. *The Atheist's Guide to Reality: Enjoying Life without Illusions*. 1. publ. as Norton paperback. New York, NY: Norton, 2012.

Rothschild, Joan, ed. *Machina Ex Dea: Feminist Perspectives on Technology*. The Athene Series. New York: Pergamon Press, 1983.

Rousseau, Jean Jacques, and C. E. Vaughan. *A Lasting Peace through the Federation of Europe and the State of War*, 2009.

Scarry, Elaine. *The Body in Pain: The Making and Unmaking of the World*. New York: Oxford university press, 1987.

Schaff, Philip. *Nicene and Post-Nicene Fathers. First Series*. New York, NY: Cosimo Classics, 2007.

Schleiermacher, Friedrich, H. R. Mackintosh, and James S. Stewart. *The Christian Faith*. .Berkeley, CA: The Apocryphile Press, 2011.

Schuurman, E., and John Vriend. *Faith and Hope in Technology*. Toronto: Clements Pub., n.d.

Schwarz, Elke. *Death Machines: The Ethics of Violent Technologies*. Manchester: Manchester University Press, 2018.

———. "Technology and Moral Vacuums in Just War Theorizing." *Journal of International Political Theory*, 2018, 5–6.

Searle, John R. *The Rediscovery of the Mind*. Representation and Mind. Cambridge, Mass: MIT Press, 1992.

Seibert, Eric A. *Disturbing Divine Behavior: Troubling Old Testament Images of God*. Minneapolis: Fortress Press, 2009.

———. *The Violence of Scripture: Overcoming the Old Testament's Troubling Legacy*. Minneapolis: Fortress Press, 2012.

Shanahan, Murray. *The Technological Singularity*. The MIT Press Essential Knowledge Series. Cambridge, Massachusetts: The MIT Press, 2015.

Sharkey, Noel. "Automating Warfare: Lessons Learned from the Drones." *Journal of Law, Information, and Science*, 2011, 140–41.

Silverglate, Harvey A. *Three Felonies a Day: How the Feds Target the Innocent*. New York: Encounter Books, 2011.

Sims, Angela D. *Ethical Complications of Lynching: Ida B. Wells's Interrogation of American Terror*. 1st ed. Black Religion, Womanist Thought, Social Justice. New York: Palgrave Macmillan, 2010.

Simunovic, Dora, Nobuhiro Mifune, and Toshio Yamagishi. "Preemptive Strike: An Experimental Study of Fear-Based Aggression." *Journal of Experimental Social Psychology* 49, no. 6 (n.d.): 1120–23.

Singer, P. W. *Wired for War: The Robotics Revolution and Conflict in the Twenty-First Century*. New York: Penguin Books, 2010.

Singer, Peter. *Practical Ethics*. 3rd ed. New York: Cambridge University Press, 2011.

Future of Life Institute. "Slaughterbots Are Already Here," n.d. https://futureoflife.org/project/lethal-autonomous-weapons-systems.

Smith, Joshua K. *Robotics Persons: Our Future with Social Robots*. S.l.: Westbow Press, 2021.

Smith, Martyn John. *Divine Violence and the Christus Victor Atonement Model: God's Reluctant Use of Violence for Soteriological Ends*. Eugene, Oregon: Pickwick Publications, 2016.

Soames, Scott. *Beyond Rigidity: The Unfinished Semantic Agenda of Naming and Necessity*. Oxford; New York: Oxford University Press, 2002.

Social Robots and the Good Life: The Normative Side of Forming Emotional Bonds. S.l.: TRANSCRIPT VERLAG, 2022.

Sokol, Michelle. "Mennonite Seminary Apologizes to Victims of Famed Theologian John Howard Yoder." *National Catholic Reporter*, April 9, 2015. https://www.ncronline.org/news/accountability/mennonite-seminary-apologizes-victims-famed-theologian-john-howard-yoder.

Stassen, Glen. "Just Peacemaking as Hermeneutical Key: The Need for International Cooperation in Preventing Terrorism." *Journal of the Society of Christian Ethics* 24, no. 2 (2004): 171–91.

Stern, David. "The Captive Woman: Hellenization, Greco-Roman Erotic Narrative, and Rabbinic Literature." *Poetics Today* 19, no. 1 (n.d.): 91–127.

Fmpr. "Strong Protest by the Chinese Government Against the Bombing by the US-Led NATO on the Chinese Embassy in the Federal Yugoslavia," n.d.

ee.china-embassy.gov.cn. "Strong Protest by the Chinese Government Against the Bombing by the US-Led NATO on the Chinese Embassy in the Federal Yugoslavia," n.d. http://ee.china-embassy.gov.cn/eng/jbwzlm/zggk/xzgwjjs/200405/t20040517_2400547.htm.

Suchocki, Marjorie. *The Fall to Violence: Original Sin in Relational Theology*. New York: Continuum, 1995.

Swamidass, S. Joshua. *The Genealogical Adam & Eve: The Surprising Science of Universal Ancestry*. Downers Grove, Illinois: IVP Academic, 2019.

Terry, Jennifer. *Attachments to War: Biomedical Logics and Violence in Twenty-First Century America*. Next Wave: New Directions in Women's Studies. Durham: Duke University Press, 2017.

Tether, T. "Welcoming Speech," 2002. http://archive.darpa.mil/DARPATech2002/presentations/diro_pdf/speeches/TETHER.pdf.

The Ante-Nicene Fathers. 3: Latin Christianity - Its Founder Tertullian; Apologetic, 2. Anti-Marcion, 3. Ethical. Reprinted. Grand Rapids, Mich: Eerdmans, 1986.

Thomas. *Summa Theologica*. Complete English ed. Westminster, Md: Christian Classics, 1981.

Townes, Emilie Maureen, and Angela D. Sims, eds. *Womanist Theological Ethics: A Reader*. 1st ed. Library of Theological Ethics. Louisville, Ky: Westminster John Knox Press, 2011.

Tsevat, M. "The Basic Meaning of the Sabbath." *Zeitschrift Für Die Alttestamentliche Wissenschaft* 84 (1972): 447–59.

Congress.gov. "Uniting and Strengthening America by Providing Appropriate Tools Required to Intercept and Obstruct Terrorism (USA Patriot Act) Act of 2001," 2001. www.congress.gov/107/plaws/publ56/PLAW-107publ56.pdf.

"U.S. Department of Defense Directive 3000.09: Autonomy in Weapon Systems." U.S. Department of Defense, 2012.

Vaux, Roland de. *Ancient Israel: Its Life and Institutions*. The Biblical Resource Series. Grand Rapids: Livonia, Mich: W.B. Eerdmans Pub. Co. ; Dove Booksellers, 1997.

Veliz, Carissa. "LaMDA Is Nothing Like a Person: This Is Why." *The Wire*, June 22, 2022. https://science.thewire.in/the-sciences/why-lamda-nothing-like-a-person.

Verdoux, Philippe. "Transhumanism, Progress, and the Future." *Journal of Evolution and Technology* 20, no. 2 (2009): 50.

"Violence Against the Elderly Has Risen During COVID, UN Expert Warns." United Nations, June 14, 2021. https://news.un.org/en/story/2021/06/1093962.

Walker, Alice. *In Search of Our Mothers' Gardens: Womanist Prose*. A Harvest Book. Orlando: Harcourt, 2004.

Waller, James. *Becoming Evil: How Ordinary People Commit Genocide and Mass Killing*. 2nd ed. Oxford; New York: Oxford University Press, 2007.

Wassermann, Christoph. *Struktur Und Ereignis: Inter Disziplinäre Studien in Physik, Theologie, Und Philosophie.* Faculté-autonome de theologie protestante, 1991.

Webb, William J. *Corporal Punishment in the Bible: A Redemptive-Movement Hermeneutic for Troubling Texts.* Downers Grove, Ill: IVP Academic, 2011.

Webb, William J., and Gordon K. Oeste. *Bloody, Brutal, and Barbaric? Wrestling with Troubling War Texts.* Downers Grove, Illinois: IVP Academic, an imprint of InterVarsity Press, 2019.

Weinberger, Sharon. *The Imagineers of War: The Untold Story of DARPA, the Pentagon Agency That Changed the World.* First Vintage books edition. New York: Vintage Books, a division of Penguin Random House LLC, 2018.

Weinfeld, W. M. "Sabbath, Temple Building, and the Enthronement of God." *Beth Mikra* 69 (1977): 188–93.

Weitzman, Michael. *The Syriac Version of the Old Testament: An Introduction.* Cambridge, MA: Cambridge University Press, 1999.

Wenham, Gordon. *Genesis 1–15.* Vol. 1. World Biblical Commentary. Waco, TX: Word, 1987.

West, S. "The Lex Talionis in the Torah." *Jewish Bible Quarterly* 21 (1993): 183–88.

West, Traci C. *Wounds of the Spirit: Black Women, Violence, and Resistance Ethics.* New York: New York University Press, 1999.

Westbrook, Raymond. *A History of Ancient Near Eastern Law.* Atlanta, GA: SBL Press, 2017.

Wienhues, Anna. "Respecting the Nonhuman Other: Individual Natural Otherness and the Case for Incommensurability of Moral Standing." *Environmental Values* 30, no. 1 (2021): 1–20.

Williams, Delores S. *Sisters in the Wilderness: The Challenge of Womanist God-Talk.* Maryknoll, New York: Orbis Books, 2013.

Williams, James G. *The Bible, Violence, and the Sacred: Liberation from the Myth of Sanctioned Violence.* 1. ed. San Francisco: Harper, 1991.

Wilson, James Q. *The Moral Sense.* 1. Free Press Paperbacks ed. New York, NY: Free Press Paperback, 1997.

Wink, Walter. *Jesus and Nonviolence: A Third Way.* Facets. Minneapolis, MN: Fortress, 2003.

———. *Jesus and Nonviolence: A Third Way.* Facets. Minneapolis, MN: Fortress, 2003.

Wright, N. T. *Christian Origins and the Question of God. Vol. 3: The Resurrection of the Son of God.* Nachdr. Vol. 3. Minneapolis: Fortress Press, 2008.

Wuellner, Bernard J. *Summary of Scholastic Principles.* Whitefish, Montana: Literary Licensing, 2013.

Young, Andrew. *An Easy Burden: The Civil Rights Movement and the Transformation of America.* Waco, TX: Baylor University Press, 2008.

živadinovič, Dojčin. "The Origins and the Antecedents of Joachim of Fiore's (1135–1202): His Historical -Continuous Method of Prophetic Interpretation." *Andrew University Seminary Studies* 56, no. 2 (2018): 330–31.

Žižek, Slavoj. *Violence: Six Sideways Reflections.* 1st Picador ed. Big Ideas/Small Books. New York: Picador, 2008.

FSC
www.fsc.org
MIX
Papier | Fördert gute Waldnutzung
FSC® C083411

Kolibri 360 GmbH
Ferdinand-Jühlke-Straße 7
99095 Erfurt, Deutschland
produktsicherheit@kolibri360.de